LECTURE NOTES
——ON——
SOLUTION CHEMISTRY

LECTURE NOTES
—— ON ——
SOLUTION CHEMISTRY

Viktor Gutmann
Gerhard Resch

Wien, Technical University, Austria

World Scientific
Singapore • New Jersey • London • Hong Kong

Published by

World Scientific Publishing Co. Pte. Ltd.

P O Box 128, Farrer Road, Singapore 9128

USA office: Suite 1B, 1060 Main Street, River Edge, NJ 07661

UK office: 57 Shelton Street, Covent Garden, London WC2H 9HE

Library of Congress Cataloging-in-Publication Data

Gutmann, Viktor.
    Lecture notes on solution chemistry / by Viktor Gutmann and
Gerhard Resch.
       p.   cm.
    Includes index.
    ISBN 9810222580
    1. Solution (Chemistry)   I. Resch, Gerhard, 1937–   . II. Title.
QD541.G86   1995
541.3'4--dc20   95-24203
                                          CIP

CHEMISTRY

British Library Cataloguing-in-Publication Data
A catalogue record of this book is available from the British Library.

Printed in Singapore.

PREFACE

One of the authors of this book has been active in research concerning liquid solutions for nearly half a century [1] and both of them have been in close cooperation for more than 20 years on this subject as well as on the philosophical foundations of the acquisition of knowledge of "things as they are". These activities have been linked with extensive lecturing. The arrangement of the 21 lectures in this book should help to provide a better understanding of the liquid state, particularly of water and its solutions. It is not a new theory, derived from imagination that is presented, but rather a new approach to the understanding of the qualities of solutions.

In the first Chapter a brief historical account about the development of solution chemistry is presented, together with its present emphasis on quantitative aspects. The second Chapter deals with the fact that atoms and molecules are real only within the continuous relationships found in nature, from which they actually cannot be removed as "isolated entities".

Molecular changes are described by the extended donor-acceptor approach, which is independent of model assumptions. Its limitations for an understanding of the properties of a liquid lead us to turn our attention to water, the most abundant and versatile liquid on earth and a "conditio sine qua non" for all life processes.

In order to learn about the "potentialities" i.e. the abilities of water and its molecules which are developing under appropriate complex conditions, it is necessary to study water under complex conditions, not only under selected and simplified experimental conditions.

It is pointed out that a solution always acts as a unity and this requires a high differentiation in itself. Chapters 6 to 12 are devoted to a description of various observations on water and its solutions, both under natural and under experimental conditions, and to the elucidation of their molecular differentiation. In Chapters 13 to 15 we present the current state of knowledge of chemistry in non-aqueous solutions and their differentiations.

After consideration of all of these facts, Chapter 16 is devoted to the methodology of ordering the observed facts according to natural requirements rather than according to our model assumptions. The differences and the connections between quality and quantity are outlined and the hierarchical differences between different molecules established. These observable differences lead to the conclusion that each observable thing must be subject to a "system organization", which is a

---

[1] H. Gamsjäger, Coord. Chem. Rev. **135** (1994) 1.

requirement for its characteristic quality. Their study requires similarity considerations rather than the exclusive consideration of quantitative aspects that is applied in system theory.

In the following Chapters the approach is presented as a means of gaining knowledge of the system organization of liquid water and its solutions. It is shown that supercooled water is highly organized and that water provides for both the unity of the human body as well as its differentiation. Chapter 20 deals with the organization of non-aqueous solutions with emphasis on the role of traces of water. In the last Chapters guidelines are presented for intramolecular system organizations. In these ways a new outlook appears to be forged, not only for chemistry, but also for biochemistry, biology, medicine and other branches of science.

It is hoped that extended research in all fields of science may be stimulated for scientists who are prepared to overcome their specialist attitude and to reorganize their knowledge in order to learn more about the natural requirements for the existence of things, their qualities, and their system organizations.

The authors wish to express their special gratitude to Prof. Reginald F. Jameson, University of Dundee/Scotland for reading the manuscript and for providing useful suggestions, to Dr. Christiana Kuttenberg for the preparation of the tables and to Mr. Harald Schauer for the production of all of the figures. Special thanks are due to the daughter of the first author, Elisabeth Bruckner, for the preparation of the "camera ready" layout and last, but not least, to Dr. Rumen Duhlev of World Scientific Publishing Co. for having encouraged us to make this contribution and for its publication.

Vienna, May 1995 Viktor Gutmann and Gerhard Resch

CONTENTS

CHAPTER 1

DEVELOPMENT AND PRESENT STATE

1. The Qualitative Approach

We come into contact with nature by means of our senses, which provide highly specialized "gateways" to the natural objects. Each object has a faculty to influence the environment in highly specific ways. The actual changes, which are induced in our senses, are called qualities. Each quality is an expression of the object as it is perceived. The senses are set in motion by the objects and this guarantees the adequate representation of the objects. The problem of sensual perception is, however, not within the scope of this series of lecture notes.

The ability of quality to structure the environment requires its own characteristic structure. This means that quality has a form which cannot be transmitted without motions. For this reason quality is defined in the broadest sense as "form in motion". Quality is always "happening" and hence it cannot be described by a rigid model, as is possible for quantity. This feature of quality makes it so difficult to define quality in an abstract manner, so that *Thomas Aquinas* referred to quality as "res obscura et varia". Further difficulties are provided by the impossibility to measure and to manipulate a quality itself (see p. 2 f and p. 169 ff).

On the other hand it should be kept in mind that a quality is a prerequisite for a quantity to be measured and hence it is virtually impossible to exclude quality from scientific investigations.

Chemistry is concerned with the qualities of materials and their changes and it has been developed by almost exclusive considerations of qualitative aspects up to the second half of the 18th century. The unique properties of liquid water have been realized in early times. Many materials were found soluble in this liquid and many changes observed in its solutions, for example the fermentation of fruit juices to wine and finally to vinegar. Mention has been made about its neutralization in the bible in the proverbs of *Solomo*, **25**, 20.

An essential step towards the metal age was the observation that metals can be melted and cast into certain forms, and that for their extraction from the ores melting procedures were important. *Agricola* (1494 – 1555) collected the enormous knowledge of his time on mining and metallurgical practise and he is considered among the first to find a natural science upon observation. His main work "De re metallica" [1] was translated into English in 1912 by the mining engineer *Hoover* (later US president), who regarded *Agricola* as the originator of the experimental approach of the sciences, "the first to find any of the natural sciences upon research and observation".

The alchemists attempted to discover the relationships of man to the terrestrial nature of the cosmos and to apply the relationships to the benefit of mankind. In these ways alchemy may be considered as an art to achieve transmutations, i.e changes in material qualities and to learn about these changes. Sulphur, the "stone that burns" (a clear statement of its quality) was crucial to alchemy just as mercury, the liquid metal, which had been known since the 3rd century BC. Mercury was found to dissolve most other metals to form amalgams. The red powder formed by its interaction with sulphur was also found as a mineral in nature which was converted into the metal by strong heating in air. The manipulation of certain virtriols led to the production of sulphuric acid, which gave nitric and hydrochloric acid by reacting with salpetre and rock salt respectively. Acids were characterized by the following similarities: sour taste, ability to change the colour of certain plant dyes to red, precipitation of sulphur from its alkaline solutions and loss of all of these properties by the interaction with bases. It may be emphasized that the enormous amount of knowledge about changes in qualities obtained by alchemists provided the background for the further development of chemistry and of modern chemical technology.

The Phlogiston theory was advanced by *Stahl* in 1700, and this had a great influence on chemical thinking up to the end of the 18th century. Changes in quality by burning metals were considered as due to the release of phlogiston and represented by the equation:

$$\text{"metal"} \Leftrightarrow \text{"calx"} + \text{"phlogiston"}$$

Schwarzenbach [2] has pointed out that in this way the processes of combustion, breathing, deterioration and fermentation were presented in a unified way and that phlogiston was not considered a substance but rather a "principle". It stands for something that is lost in the course of an oxidation and for the "principle" which is immanent in the metals. He referred to the fact that the "electron gas" is characteristic for the metallic state and that the release of electrons in the course of an oxidation is now formulated in formal analogy to that provided by the equation of the phlogiston theory:

$$M = M^+ + e^-$$

This general formulation has been lost by the idea of *Lavoisier* to regard the presence of oxygen as necessary for an oxidation reaction and also for an acid. The argument that hydrochloric acid does not contain oxygen, was dismissed by introducing the hypothesis that chlorine was not an element but rather a combination of an element with oxygen ("oxymuriaticum").

2. Advancement of Quantitative Considerations

Lavoisier is frequently considered to have introduced the quantitative methods into chemistry, but other chemists had initiated quantitative measurements in the middle of the 18th century (see Chapter 3). Quantitative methods provided the basis for the development of modern chemistry, but in classical physics this approach had been applied much earlier. *Copernicus* (1473 – 1543) had not claimed to have made a new discovery, but simply to have found a new way for mathematical description of the motions of the planets. He expressed this in the preface to his work "De revolutionibus orbium coelestium" as follows: "It is not necessary for the hypothesis to be true, or even close to the truth; as long as it allows calculations that correspond to our observations, it is enough".

The interest in quantitative aspects is not only due to their measurabilities, but also to the possibility of modifying the world. This cannot be achieved by acting on qualities, because these are indivisible and cannot be modified by man. *We can only observe changes in qualities due to changes in local positions of the things.* For example, a neutralization reaction occurs after uniting an acidic and an alkaline solution, but it would be wrong to say that this reaction has been carried out by us. The changes in qualities occur according to the natural laws, on which we have no influence. On the other hand, we can act on quantity and hence modify the physical world in this way.

For *Galilei* (1564 – 1642) the book of nature was written in mathematical formulas. He gave the direction "measure whatever can be measured and render measurable, what cannot yet be measured". This instruction, however, does not consider the fact that quality itself cannot be measured, nor can it be made measurable. Thus, in this instruction, a quality is not comprehended and it provides no advice as to the attitude towards those phenomena that are unmeasurable in principle. The argument that is sometimes made that qualities are measurable is due to the confusion of quality as a requirement for quantitative measurement with the result of the measurement.

Soon research on quantities was considered as indispensable in science and knowledge of qualities as inferior. This may be illustrated by the statement made by the German philosopher *Kant* (1724–1804): "I state that chemistry is not more than a systematic art and collection of descriptions of experiments, which never can reach the status of a real science, because the principles of chemical changes cannot be expressed by mathematics.

For a quantitative mathematical description a clear definition of the objects of research is required. These are, however, always part of a complex whole, within which they cannot be precisely defined. In order to be able to do this one would have to extract them from their natural context. This, however, would mean the loss of any possibility to observe and measure them. *Stapp* [3] has expressed this dilemma in the following way: *"The observed system is required to be isolated in order to be defined, yet interacting in order to be observed".*

Consequently, the requirements for mathematical description have to be artificially constructed by "mentally" detaching the objects from their context. With this, however, it is no longer reality which is being investigated, but the idealized models of the objects in question.

This development has been supported by the *analytical method*. Its success has been immense. However, these gains have to be bought at the expense of losses of information about the mutual relations between the parts within the natural conditions. By the detachment of the parts from the latter, the characteristics of the parts within the whole are irretrievably lost, but they are indispensable for the functions within the superordinated system [4].

When chemists became interested in quantitative relationships, they did not try to replace the search for qualities by that for quantities but rather to proceed jointly with qualitative *and* with quantitative studies. The establishment of the periodic table serves as an example for the success of these joint considerations [5]. *Doebereiner* had noted similarities in properties between certain elements and grouped them together in "triads". *Mendeleev* arranged the elements according to their quantitative characterization by increasing atomic weights and grouped them by making use of similarity considerations, as they are required in order to gain an understanding of qualities. In this way he was even able to predict properties of elements hitherto undiscovered.

3. Reversible Thermodynamics

This approach was successfully applied in the early stages of the development of thermodynamics. The first law of thermodynamics is based on various qualitative observations by *Mayer*, a doctor of medicine in Germany and by *Joule* in England in the decade of the 1840s. The original papers of each of them were rejected by editors in Germany and England respectively. *Mayer* had realized the need for finding quantitative relationships between heat and work and carried out experiments, but the correct value for the mechanical equivalent of heat was found by *Joule*.

The second law of thermodynamics is based solely on observation, namely the impossibility of the construction of a perpetuum mobile of the second kind. The essential content of this law has been formulated by *Clausius* as follows: "A passage of heat from a colder to a hotter body cannot take place without compensation". According to results of quantitative measurements, *Clausius* found it indispensable to introduce a mathematical term in order to account for the apparent loss in energy of a system in the process of work. This mathematical term is known as *entropy*. In the course of an irreversible process such as the change of a solid to a liquid, of a liquid to a vapour, the mixing of hot and cold gases or the occurrence of chemical changes, the entropy is always found to increase. When any actual process occurs, it is impossible to invent a means of restoring every system concerned to its original condition. Any actually occurring (spontaneous) process is irreversible and *the gain in entropy is a quantitative measure of its irreversibility* [6].

The quantitative treatment of the second law required, however, the introduction of massive approximations, partly in open contrast to the statement made above, namely the assumption of a reversible process. In this case no entropy change would occur and hence the introduction of this term as a measure of the irreversibility of the actual process superfluous. The fact, that the entropy change is never zero, means that a process considered as reversible is actually irreversible. This inconsistency has been verbalised by stating that thermodynamics actually deals with systems near and not at thermodynamic equilibrium.

The precise mathematical description requires the introduction of another unrealistic idealization, namely that of a closed system. This, however, would be both unobservable and unmeasurable. Furthermore, the properties of the actual system are considered to be fully determined by the variables of state. These quantities are volume, pressure, temperature, concentration and they are considered as independent from the history of the system, from the effects at the interface and from its actual environment (which appears to have been artificially eliminated by the assumption of the closed system). These quantities are also independent of the structural differentiation of the system under consideration.

Without these restrictions the mathematical method developed by *Gibbs* in the years 1873 - 1878 would never have been so orderly and systematic as is required for the mathematical foundations of classical thermodynamics.

4. The Ionic Theory

A very important step in the development of solution chemistry has been provided by the formulation of the theory of electrolytic dissociation advanced by *Arrhenius* in his doctoral thesis presented to the University of Uppsala in 1884. Although the members of the faculty were sceptical about the theory, *Arrhenius* was awarded the doctoral degree and became a lecturer in chemistry, although not without the strong support by *Ostwald* in Germany.

With this theory the modern science of physical chemistry has been founded. It provided an entirely new look on acid-base chemistry as well as on electrochemistry and chemical kinetics. It opened up new perspectives for the development of structural considerations in solution. Solute-solvent interactions in water are explained by hydration of the solute species, but in the absence of observable borderlines between hydration sphere and bulk solution, border-lines had to be artificially introduced for quantitative considerations. In the primitive "sphere in continuum model" the hydrated ions are considered as moving freely in a continuum provided by non-affected water molecules.

Liquid water was shown to be subject to a self-ionization equilibrium. An acid was consequently defined as a solute that increases the concentration of hydrogen ions and a base as a solute that increases the concentration of hydroxide ions. By analogy, for ionic reactions in liquid ammonia the ammono-system of acids and bases was developed by *Franklin* [7]. Similar considerations were developed for many other solvents, such as sulphuric acid, acetic acid, liquid hydrogen fluoride, the alcohols and the acid amides [8]. They led to the development of the interpretation of acid-base reactions by *Brönsted* [9] and *Lowry* [10]. These definitions can be applied to all protonic solvents, since proton transfer reactions are considered as responsible for both the auto-ionisation of the amphoteric solvent molecules and the acid-base interactions in their solutions. In order to extend this concept to non-protonic solutions, the concept of ionotropism has been proposed [11]. For example, the self-ionisation of liquid bromine trifluoride and the acid-base interactions in this solvent [12] were explained by fluoride ion transfer reactions and an acid defined as a fluoride ion donor and a base as a fluoride ion acceptor respectively.

Although the parameters of the elementary electrostatic theory could not be applied to any of these solvents, they are still termed "polar" solvents as distinguished from non-polar and from metallic solvents (see p.7).

The attempts to formulate a precise mathematical treatment had, however, to be based on a number of unrealistic restrictions. In the *Debye-Hückel* theory of interionic attraction Coulombic forces are considered exclusively and this is not in agreement with the quantum-mechanical requirement for charge transfer. The ions are considered as unpolarizable point charges, each possessing a symmetrical Coulombic field, the dielectric constant as independent from the composition of the solution and the dissociation of strong electrolytes as complete [13]. Numerous improvements of this theory have been advanced, but the theory can be applied only to a small number of highly diluted solutions in water.

5. Chemical Kinetics

Many systems do not reach equilibrium conditions, although the thermodynamic conditions are favourable. The study of reaction rates revealed that a reaction may proceed via various steps which are not expressed in the chemical equation. Intermediate steps had to be postulated and their formation investigated. It was found that the so-called transition state could be reached only, if sufficient energy was provided for its formation.

This means, that even for the performance of a reaction, for which the free enthalpy is negative, energy must be provided in the early stages and this amount of energy is called activation energy. Efforts have been made in order to minimise the activation energy and in following this purpose, the branch of catalysis has been developed.

From the results of kinetic and stereochemical studies reaction mechanisms are postulated. Each of these mechanistic interpretations is subject to modifications, but classifications of reactions were possible and useful in all areas of solution chemistry, including coordination chemistry [14] and organic chemistry [15 – 17]. Mainly due to the contributions by *Ingold* [15] and *Robinson* [16] the field of physical organic chemistry was developed and reactions in solution classified as electrophilic and nucleophilic substitution reactions, as elimination and addition reactions.

More recently new experimental techniques have been developed, which permit the measurement of the rates of extremely fast reactions, namely the temperature-jump method and various stopped-flow techniques.

The path-leading contributions for the development of all of these quantitative methods in solution chemistry have been well documented by the award of the following Nobel prizes:

1901 to *H.J.van't Hoff* "for the discovery of the laws of chemical dynamics and osmotic pressure in solutions",

1903 to *S.Arrhenius* "for the services he has rendered to the advancement of chemistry by his electrolytic theory of dissociation",

1909 to *W.Ostwald* "for his work on catalysis and for his investigations into the fundamental principles governing chemical equilibria and rates of reactions".

6. Non-Aqueous Solutions

It has been mentioned that ionic reactions were found to take place in many other liquids than water and that these liquids are called "polar" solvents, although the polarity of the solvent molecules does not account for their solvent properties. Some of them have been used for a long time in organic chemistry, such as diethyl ether and more recently acetonitrile or dimethyl sulfoxide. Based on the donor-acceptor concept, empirical solvent parameters were introduced [18] and even colour indicators found which respond to the Lewis acidity and Lewis basicity respectively of the non-aqueous solution under consideration [19] (see Chapter 13).

The solvent parameters were also applied to non-polar solvents such as benzene or carbon tetrachloride. In these covalent solvents many covalent compounds are soluble. For example, iodine dissolves much more readily in benzene or in carbon tetrachloride than in water. Its solution in benzene is red, in carbon tetrachloride violet and in alcohol brown. This indicates that rather specific solute-solvent interactions take place.

It may be mentioned that none of the non-aqueous solvents can compete with the enormous versatility of liquid water and its solutions and that the term "non-aqueous" means solvents other than water and not free from water. None of these solvents can be freed from the last traces of water, which cannot be removed when present in concentrations in the ppm region (Chapter 15).

Another group of solvents is provided by the liquid metals. Liquid mercury dissolves most metals and their solutions, either liquid or solid, are known as amalgams. Most alloys are actually solid solutions of metals in metals.

According to qualitative considerations, the following similarity rules have been formulated for the solubilities:
1. Polar solvents dissolve polar solutes,
2. Non-polar solvents dissolve non-polar solutes,
3. Metallic solvents dissolve metallic solutes.

7. Solute Structures

Mainly due to the development of modern spectroscopic techniques, access to solute structures has been made possible. Remarkable advances have been made in this field around 1950 by optical spectroscopies, such as infrared, Raman, and ultraviolet spectroscopies as well as by spin spectroscopies, namely nuclear magnetic resonance and electron spin resonance spectroscopy. In addition Mössbauer spectroscopy and small angle X-ray scattering were developed. Their combined application allowed the elucidation of many solute structures, usually within a few days. New aspects were found on the structures and on the dynamics of coordinated water molecules in the hydration shells [20], but even the mean values of bond distances vary considerably under slightly different conditions. This is because it is impossible to define sharply a hydrated species in solution or to characterize it by a fixed coordination number [20].

These spectroscopic techniques and other methods allow the investigation of the influence of the solvent on structures and properties of the dissolved species, as well as on complex equilibria in solution [21,22].

8. Irreversible Thermodynamics

The concept of irreversible thermodynamics was developed within the last decades although it is based on observations which have been made in the last century. The first report on a dissipative structure has been made in 1828 and later honeycomb-like structures were found to remain after evaporation of schellack in petrolether and these are known as *Benard*-cells [23]. Non-equilibrium thermodynamics was introduced by *Onsager* [24] in 1931, but its importance for solution chemistry has not been realized at this time. In 1950 *Belousov* had discovered that a solution of citric acid in water with acidified bromate and a yellow ceric salt turned colourless and returned to yellow periodically [25]. His paper was rejected by the referees, who advised him that "his supposedly discovered discovery would merit publication only if accompanied by a demonstration that existing theory was flawed [25]". Due to the contributions of *Zhabotinsky*, this reaction became known as *Belousov - Zhabotinsky* reaction, but its mechanism is still not known in detail.

Rhythmic processes are of special significance in living systems, for example the metabolic oscillations that are found in the course of glycolysis [26]. These processes are extremely important in keeping the energy balance in cells, as adenosin triphosphate is produced as a result of the catabolism of glucose.

The same structure may be developing different new structural features under different conditions. A structure can therefore no longer be considered as completely conservative, as it develops dissipative aspects [27,28]. Each of these terms is, therefore, not characteristic for a given "isolated" system, but rather for the system under specified conditions. Hence the terms conservative and dissipative indicate "tendencies" to conservation and dissipation respectively (see Chapter 9).

9. Modern Calculation Techniques

The trends to follow the quantitative aspects has been tremendously increased by the spectacular development of modern calculation methods around 1970 [29]. They became possible owing to the development of modern electronics and high-speed electronic computers as well as to the introduction of idealizations. Apart from numerical quantum chemistry, the methods of molecular dynamics (MD), Monte Carlo (MC)-simulations and molecular mechanics (MM) calculations are applied to systems containing a number of molecules similar to that in bulk systems [20].

The basic assumption of MD-simulations [30] is that any kind of interatomic interactions can be described in terms of the sum of the interactions of atomic pairs, and thus the multibody problem is avoided. For a bulk system a periodic boundary condition is artificially introduced, long range interatomic forces are "corrected" and a set of "good" pair functions is selected. If the potential functions are "reasonable", acceptable results may be obtained, although ions are usually considered in terms of point charges and ionic radii.

In Monte Carlo (MC) simulations atomic configurations are "created randomly" and static aspects simulated. Due to this selection procedure physically irrelevant results may be obtained and hence the procedure requires distinctions to be made between "acceptable" and "unacceptable" configurations with criteria for the "acceptability" which are not always convincing [31].

10. Conclusions

22 years ago, the state of solution chemistry has been characterized by *Franks* [32] as follows: "Looking back over published work since the beginning of this century, one is struck by the enormous efforts made by chemists to document the properties of solutions. Unfortunately, one is also struck by the fact that few of the experimental data originating from before 1955 can be utilised in basic studies of solution processes on a molecular scale".

10 years later the following comment has been made by *Primas* [33]: "The molecular view has triumphed in physics, chemistry and biology with immense practical results. In the main, chemistry has fulfilled its molecular program ... Molecular theories describe some aspects of matter but it is not wise to think that they can give us a description of reality".

Molecular theory provides an excellent framework for the description of quantitative aspects involved in the course of chemical changes both on the macroscopic and on the molecular level. Whereas the quantitative features are adequately described, the qualitative aspects are not appropriately considered. The gain in precision seems, therefore to be outweighed by a loss in scope and dignity of the knowledge about the real things.

We shall, therefore, examine the question, *if the molecular concept might be suitable to provide the basis for a qualitative understanding.*

References

1. G. Agricola, *De re metallica,*libri XII, (German translation, Deutscher Hütten-buch Verlag, München 1977).
2. G. Schwarzenbach, *Chimia* **26** (1974) 101.
3. H. P. Stapp, *Phys.Revs.* **D 3** (1971) 1303.
4. V. Gutmann, *Fresenius Z. analyt. Chem.* **337** (1990) 166.
5. V. Gutmann and G. Resch, *Chemistry Int.* **10** (1988) 5.
6. G. N. Lewis and M. Randall, *Thermodynamics* (Mc Graw Hill Co, New York 1961).
7. E. C. Franklin, *The Nitrogen System of Compounds,* (Reinhold Publ. Co. New York, 1935).
8. V. Gutmann, *Quart. Revs.* **10** (1956) 451 - G. Jander, *Die Chemie in wasser-ähnlichen Lösungsmitteln,* (Springer, Berlin 1949).
9. J. N. Brönsted, *Rec. Trav. Chem.* **42** (1923) 718.
10. T. M. Lowry, *Soc. Chem. Ind.* **42** (1923) 43.

11. V. Gutmann and I. Lindqvist, *Z. physik. Chem.* **203** (1953) 250.
12. A. G. Sharpe and H. J. Emeléus, *J. chem. Soc.* (London) **1948**, 2135.
13. G. Kortüm and J. O'M. Bockris, *Textbook of Electrochemistry* (Elsevier Publ. Co. New York, 1951).
14. F. Basolo and R. G. Pearson, *Mechanisms of Inorganic Reactions*, (Wiley and Sons, New York, London, 1958).
15. C. K. Ingold, *Chem. Revs.* **15** (1934) 225.
16. R. Robinson, *Outline of an Electrochemical Theory of the Course of Organic, Reactions* (The Institute of Chemistry, London, 1932).
17. L. P. Hammet, *Physical Organic Chemistry* (Mc Graw Hill Co. New York, 1940).
18. V. Gutmann, *The Donor - Acceptor Approach to Molecular Interactions*, (Plenum Press, New York, 1977).
19. R. W. Soukup and R. Schmid, *J. Chem. Educ.* **62** (1985) 459.
20. H. Ohtaki and T. Radnai, *Chem. Revs.* **93** (1993) 1157.
21. H. L. Schläfer, *Komplexbildung in Lösung* (Springer, Berlin, Göttingen, Heidelberg, 1961).
22. M. T. Beck, *Chemistry of Complex Equilibria,* (Van Nostrand Reinhold Co., New York, 1970).
23. H. Benard, *Rev. Gen. Sci. Pure Appl.* **11** (1900) 1261.
24. L. Onsager, *Phys. Revs.* **37** (1931) 405.
25. A. T. Winfree, *J. chem. Educ.* **61** (1984) 661.
26. H. Hess and M. Markus, *Ber. Bunsenges. phys. Chem.* **89** (1985) 642.
27. P. Glansdorff and I. Prigogine, *Thermodynamic Theory of Structure, Stability and Fluctuations* (Wiley - Intersci. London, 1971).
28. I. Prigogine, *Introduction to the Thermodynamics of Irreversible Processes* (Wiley, New York, 1967).
29. U. Müller - Herold, *Chimia* **39** (1985) 3.
30. K. Heinzinger and G. Pálinkás, in *Interactions of Water in Ionic and Nonionic Hydrates* (Springer, Berlin 1987).
31. N. Metropolis, A. W. Rosenbluth, M. N. Rosenbluth, A. H. Teller and E. Teller, *J. chem. Phys.* **21** (1953) 1087.
32. F. Franks, *Water, a Comprehensive Treatiese,* Vol. 2, p.1 (Plenum Press, New York, London, 1973).
33. H. Primas, *Chimia,* **36** (1982) 293.

CHAPTER 2

ATOMS AND MOLECULES

1. Early Views

The view that unity could definitely be found behind all multiplicity has been expressed by *Anaximander* (610 - 547 BC) as *"apeiron"* (the "unlimited"). The Eleatic school of Greek philosophy, founded by *Parmenides* (born 515 BC) also considered *matter as a continuum filling completely the universe* and this view was supported by *Aristotle* (384 - 322 BC).

The second view, the "Atomic Hypothesis" is due to *Leucippos* (about 500 BC) and *Democritus* (460 - 370 BC), who considered *matter as consisting of small indivisible and unchangeable parts, the atoms.* In order to account for their unchangeability it was necessary to consider them as non-interacting entities, to avoid the idea of their continuous relationships and to assume, instead, empty space between them.

In the book "Nature and the Greeks" *Schrödinger* [1] pointed out that *Democritus* himself was aware of the fundamental difficulties with this thesis. He described a dialogue between intellect and senses. The former says: "Ostensibly there is colour, ostensibly sweetness and ostensibly bitterness, in truth there are but atoms and voids", to which the senses reply: "Wretched mind, from us you are taking the evidence, by which you would overthrow us, your victory is your own fall."

2. Ether Concept and Atomism

No further advances have been made with regard to these questions until the 14th century, when for the continuous relationships the existence of an extremely thin medium, the so-called "ether" was proposed. This view was accepted by *Copernicus* (1473 - 1543) as a requirement for the motion of the earth, which he could no longer consider as an inertial system at rest. *Descartes* (1596 - 1650) denied also the existence of a vacuum and opposed the attempt made by *Gassendi* (1592 - 1655) to revive atomism.

The ether concept was also accepted by the founder of the wave theory of light, *Huygens* (1621 - 1695), who considered a medium as necessary for the support of the waves of light. *Leibniz* (1646 - 1716) was convinced of a "prestabilized harmony" in all nature and supported the *Aristotelian* view that it was meaningless to talk about the motion of a body within an empty space. The development of the theory of gravitation is due to *Newton* (1642 - 1727), who considered the interactions of bodies as a function of their distance. He supposed the body in question to act everywhere in space and hence he saw no reason for a sharp distinction between empty and occupied space, but assumed the presence of attractive forces between all massive bodies for the gravitation.

Based on this approach *Boscovich* (1711 - 1787) considered the atoms as dynamic centres of force and suggested to account for all known physical effects in terms of actions at a distance between atomic "point particles".

The investigation of the quantitative aspects of chemical changes was begun by *Lomonossov* (1711 - 1765) and extended both by *Richter* (1762 - 1807) and *Proust* (1754 - 1826). *Dalton* (1766 - 1844) developed the "Atomic Theory" by formulating the laws of constant and multiple proportions. He assumed the chemical elements to consist of indivisible atoms and suggested to characterize each element by the weight of its atoms. In this way he expressed quantitative aspects and provided the pathway for the development of modern chemistry. The determination of the atomic weights required the preparation of highly purified materials, the performance of complete chemical conversions and the precise analytical characterization of both reactants and reaction products.

3. The Field Concept and the Ether Concept

Despite the great success of the atomic concept, *Faraday* (1791 - 1867) was motivated by a belief in the unity of nature. In 1820 he entered into his note book [2]: "Observing more closely the relationships between all of the forces, one cannot say that one moves the other, but only that they are all being moved by a common factor". He was convinced that for all forces a continuous, supporting medium was necessary. In 1838 he had drawn lines of force from the alignment of iron cuttings caused by the force of a magnet. *By means of these "lines of force" he characterized the magnetic field.* This field concept is the physical demonstration of the continuous relationships between the concrete particles.

However, this concept was not accepted by his contemporaries for nearly 20 years. The mathematician *Airy* called it a "vague affair" and the physicist *Tyndall* even blamed *Faraday* for confusing scientists. When *Maxwell* (1831 - 1871) published the paper entitled "On Faraday's Lines of Force" in 1857, the field concept provided the background for the electromagnetic theory, first fully set out in 1864. His theory is a theory of *waves in a continuous medium*. This theory seemed, therefore to express characteristics of the ether and required the existence of tensions in order to allow for the oscillations, which were shown to be propagating with the velocity of light.

The ether provided also the background for *Fresnel* (1788 - 1827) in developing a complete theory of light propagation. For this elastic medium the following properties were assumed: weightless, transparent, frictionless, literally permeating all matter and space, although sensually undetectable. The physical properties of the ether were described mathematically and the atoms considered to move in this medium, which was considered to "penetrate" all bodies.

The ether concept was therefore appreciated and the field concept well established, when *Loschmidt* (1821 - 1895) presented his paper on the "Constitution of the Ether" [3]. He was fully convinced of the atomic concept to which he had contributed enormously by the assessment of the number of atoms contained in one mole, now called *Avogadro* constant, but convinced of the necessity of a medium connecting the atoms.

According to classical physics, the ether should provide a suitable reference system for the speed of light, as the ether was thought to make up the basic substratum of the universe. In 1887 *Michelson* (1852 - 1931) performed an experiment in order to measure the velocity of the earth against the ether. However, the experiment showed that there was apparently no motion of the earth relative to the ether. In order to explain this result, the ether hypothesis was abandoned, although the experiment had only shown that the speed of the ether was not synchronous with that of the earth, but rather with that of the light. It showed, that the premises of classical mechanics cannot be applied to the dynamically maintained continuum conditions and that it was impossible to draw any conclusions about the motion of the reference system.

Although it seemed no longer necessary for the interpretation of the speed of light to maintain the ether concept, the transmission of light still requires the presence of a continuous medium and this demand is also a requirement for the theory of relativity. The special theory of relativity led *Einstein* (1879 - 1955) to the concept of the gravitational field, which expresses the continuous relationships between bodies.

4. Quantum Mechanics

Continuous relationships are also required for quantum mechanics, although this theory has been designed originally so as to provide a broader field of application for *Newton's* theory, namely to describe the behaviour of radiation from an energy distribution stand point. *Planck* (1860 - 1944) assumed submicroscopic units as causing the radiation, which he called oscillators [4].

The frequency of the oscillator is increased by adsorption of energy (E) and frequency (v) of the oscillator : $E = h.v$. This equation demands that for energy differences approaching zero ($E \to 0$), either v or h must approach zero. As long as a change in frequency is measurable, the frequency cannot be zero. *Planck* has realized that the constant h must also have a finite value. So he concluded that E is to have a finite value. This result from mathematical considerations led him to the entirely new suggestion that an oscillator is unable to adsorb or emit energy continuously. It only does so in integral multiples of a definite amount, depending on the frequency of the oscillator. Hence, the energy of a body is not continuously variable, but can consist only of a definite number of "quanta". This led to the subsequent conclusion that oscillators are atoms in which electrons are permitted to occupy only a discrete number of energy states. With regard to quantum theory and continuum, *Einstein* wrote in a letter to *Born* on 27. January 1920: "I cannot believe that the quanta problem can be solved by abandonment of the continuum. By analogy, one might have been tempted to save general relativity by abandoning the coordination system ... I would not know how to describe the relative motion of points without continuum." [5].

An interesting point is the fact that *Planck* himself was one of the conservative scientists of the classical epoch, who did not even accept the probability interpretation of thermodynamics. He applied the statistical method only when he saw no other possibility of providing a theoretical basis for the correct radiation formula which he had conceived intuitively.

With regard to the present context it should be emphasized that the *quantification results from the impossibility to observe or to measure continuous changes in properties. The constant h, known as Planck-constant, is considered a fundamental constant, in that it marks the demarcation line below which no measurements are feasible because of the lack of borderlines in the continuum. Thus, quantum theory requires continuous relationships, although - as a scientific theory - it is to describe discontinuities, as they are required for measurements.*

In 1924 *De Broglie* had the idea to consider propagating electrons as waves and to describe in this way the continuous propagations of electrons. *Schrödinger* considered the electrons in stable states as "structures of standing waves" and developed the well-known Ψ equation, in which the wave function Ψ, also called orbital, is a mathematical term. In his theory, he did not explicitly state that this required a continuum, but he was reminded of this in a letter by *London* [6], who wrote to him: "Four years ago you demonstrated that one has no other choice that the whole world of atoms represents a process in continuum without any identifiable fixed point. You held the truth in your hands and, like a priest, kept it as a secret".

In 1927 the most brilliant physicists of their time met in Brussels in order to discuss the consequences of the results of quantum mechanics. The result of this conference is known as "Copenhagen Interpretation", obviously so-called to appreciate the influence of the Danish physicist *Bohr*. One of the conclusions concerns the so-called "wave-particle complimentary". Both continuum - and particle concept, though mutually exclusive, are considered necessary for the appropriate description of matter. Each of these concepts describes different aspects of matter and both of them are necessary for the description of all aspects.

It has been pointed out [7] that the unspecified question whether matter consists of parts or whether it represents an undivided continuum, is based on a misunderstanding. As soon as questions are raised with regard to the quantitative characterization of a material, the particle description is necessary. If, however, an understanding of the events in their natural relations is sought (which is considered impossible according to the Copenhagen interpretation) this cannot be reached on the basis of the particle concept alone without first considering the continuum aspects of matter.

The answer to the question, whether the continuum or whether the parts are to be considered as primary, depends on the motivation of the investigator. *With regard to the recognizability, the parts are appropriated the prime position, because discontinuities are required for observation and measurement*, whereas the continuous cannot be perceived and for this reason it cannot provide the starting point for the investigation. However, with regard to the acquisition of knowledge of nature as it is (which again is beyond the scope of quantum mechanics according to the Copenhagen interpretation), it must be borne in mind that *the continuum is a prerequisite, a conditio sine qua non, both for the existence and for the observation of all things and hence the continuum is primary.* It includes the parts but the parts could not exist in its absence.

Because the continuum and the parts require each other they cannot be considered as antithetical. If we wish to obtain knowledge of nature as it is, it is *imperative to investigate the perceivable parts within and not outside the natural relationships.*

5. The Concept of the Vacuum Field

The enormous role of the continuum has been recently demonstrated by *Puthoff* [8]. He refers to the continuum as a *"vacuum field that is not empty, but full of energy"*. It acts as a dynamical background by which the states of all parts and their interactions are determined. The most fundamental concept is that *the vacuum field is fluctuating at zero- point energy level* and that the zero-point energy is so completely in balance that under ordinary circumstances its effects are unobservable.

Physical evidence is provided, for example, by the *Lamb*-shift, i.e. the departure from theory of the actual frequencies of light emitted from the electron of an excited hydrogen atom. These departures are due to the assumption for the calculation that the atom is not in a void, but, when the effects of the electromagnetic zero-point energy on the electron (from the continuous vacuum field) are taken into account, the differences are nearly eliminated [9].

Further support for the actions of the continuum are provided by the consideration of the so-called *Casimir*-effect [10]. When two parallel conducting plates are placed close to each other within a distance of 10^{-6} m, they are pushed together due to the radiation pressure from outside. These plates may, however, be pushed apart due to the radiation pressure associated with the zero-point energy of the vacuum field [11].

Puthoff [8] has further shown that within the stochastic electrodynamic formulation at the level of *Bohr*-theory, the ground state of the hydrogen atom can be precisely defined as resulting from a dynamic equilibrium between radiation emitted due to acceleration of the electron in its ground state orbit and radiation absorbed from the zero-point fluctuations of the background vacuum electromagnetic field. In this way the issue of the radiative collapse of the *Bohr*-atom is resolved simply by dropping the unrealistic assumption of an isolated atom and by considering the fluctuations of the continuous background electromagnetic field.

Likewise, the spontaneous light emission of atoms and the non-reproducibility of the values for the relaxation time of the excited state of the hydrogen atom are regarded as due to the fluctuations within the continuum [9].

Atoms and molecules are continuously interacting with the vacuum fluctuations, which are continually absorbed and re-emitted by them. This means that *atoms and molecules owe their existence and observability to their continuous interactions with the background fluctuations* [11]. *Puthoff* [8] concluded his paper. "This carries with it the attendant implication that *the stability of matter itself is largely mediated by zero-point fluctuation phenomena* in the manner described here, a concept that transcends the usual interpretation of the role and significance of zero-point fluctuations of the vacuum magnetic field".

6. Observability and Measurability of Atoms and Molecules

Sensual perceptions and measurements require the presence of border line areas. Modern science has developed techniques by which the limitations of sensual perceptions are extended. A piece of metal can be sensually perceived, as it appears separated from its environment. For the naked eye, the piece of metal appears homogeneous, but observation by means of a microscope (as an extension of the faculty of the eye) reveals that it contains many small grains separated from each other and connected to each other by the so-called "grain boundaries". When the metal is observed by means of an electron microscope, new borderlines become apparent, namely dislocations and in some cases even point-defects. When X-rays are directed to the metal, its interactions with the electron shells lead to the production of a characteristic diffraction pattern, from which the crystal structure may be derived.

The question may be raised, in what ways do the electron shells provide border lines in order to allow the measurements? It is well-known that the electron shells represent areas of low density both in mass and in negative electric charge and that they are related to the atomic nuclei, which represent areas of high densities both in mass and in positive electric charge. Their relations to the continuum cannot be ignored and these provide areas of unmeasurably small densities in mass and in electric charge and they are usually referred to as "intermolecular space".

As all of them must be considered as belonging to the continuum, the *electron shells appear indeed to provide border line areas between nuclei and intermolecular space*. These border line areas are characterized by enormously high gradients in mass and in electric charge and they represent within the continuum the discontinuities as required for observation and measurement [12,13]. Because the highly condensed areas are completely surrounded by the measurable discontinuities of the electron shells, the former are rightly considered as "nuclei" of units, named atoms and hence the term *"atomic nuclei"* is justified.

A molecule may be defined within the continuum as a subsystem, in which the electronic boundary regions of two or more atoms penetrate each other, so that the said atoms appear to be united within one observable and measurable boundary region [13]. The separation of an atom or of a molecule from the continuum is only possible mentally. An electrically neutral atom is theoretically removed from the continuum by taking from the electron shell surrounding the nucleus an amount of negative charge that equals the positive charge of the nucleus. If we remove less negative charge from the shell than there is positive charge in the nucleus, the result is called a cation. If we remove more negative charge, we get an anion. However, *isolated electrons, atoms, ions or molecules are the result of abstractions*, namely after their mental dismemberment from the continuum, by which they actually would loose their border lines and could no longer be observed.

CONTINUUM

Undivided and dynamically ordered connection, all-embracing and highly differentiated;

Realities, mutually required for existence, observability and understanding of all things;

Mutual interactions between all regions.

SUBINTELLIGIBLE	INTELLIGIBLE	SUBINTELLIGIBLE
Indirect access	Directly observable and measurable	Indirect access
Not differentiable "Supporting"	Differentiable "Separating Connections"	Not differentiable "Supporting"
HIGHLY DILUTED	BOUNDARIES	HIGHLY CONDENSED
INTERATOMIC	ELECTRON SHELLS	ATOMIC NUCLEI

SUBINDIVIDUAL REGION	INDIVIDUAL REGION (ATOM)
	Atoms are artificially "made" by considering the subintelligible highly condensed areas of the continuum as "inner parts" of the intelligible boundaries between interatomic regions and atomic "nuclei".
NOT MEASURABLE NO THEORY High flexibility QUALITATIVE Static aspects not observable	MEASURABLE QUANTUM MECHANICS Low flexibility QUANTITATIVE Static aspects observable

Thus, atoms and molecules represent observable discontinuities within the continuous relationships. Because the continuum is fluctuating, the border lines are neither sharp nor rigid. *Atoms and molecules are fictions, when considered as isolated, but realities within the continuous relationships.* This has been indicated by *Whyte* [14]: "Most philosophers and scientists have agreed that, to account for what is observed, it is necessary to infer entities, which are not observed. An unseen universe is necessary to explain the seen".

The enormous difficulties in gaining such knowledge have been formulated by *Heisenberg* [15] as follows: "For Christoph Columbus it must have been the most difficult decision to leave all known land and to sail so far towards West, that the storage on board would not allow him to return. In a similar way, completely new land in science cannot be discovered, unless one is prepared to leave at a certain point the foundations on which traditional science is based and to try to 'jump into emptiness'".

References

1. E. Schrödinger, *Nature and the Greeks,* (Sherman Lectures, Univ. College London, 1948).
2. A. Hermann, *Weltreich der Physik, von Galilei bis Heisenberg,* (Ullstein Sachbuch, 1983).
3. J. Loschmidt, *Zur Konstitution des Äthers,* (Carl Gerold's Sohn, Wien, 1862).
4. R. Sexl and H. K. Schmidt, *Raum-Zeit-Relativität,* p.182, (Vieweg, Braunschweig, 1978).
5. *Aus dem Briefwechsel Einsteins mit Max und Hedwig Born* (Vieweg, Braunschweig, 1966).
6. W. Moore, *Schrödinger, Life and Thought,* p. 148 (Cambridge Univ. Press, 1992).
7. G. Resch and V. Gutmann, *Scientific Foundations of Homeopathy,* (Barthel Publ. Germany, 1987).
8. H. E. Puthoff, *Phys. Revs.* **D 35** (1987) 3266.
9. H. E. Puthoff, *Phys. Revs.* **A 40** (1989) 4857.
10. H. B. G. Casimir, *Physica* **19** (1953) 846.
11 .P. W. Milloni, R. J. Cook and M. E. Goggin, *Phys. Revs.* **A 38** (1988) 1621.
12. V. Gutmann and E. Hengge, *Anorganische Chemie,* 5. Auflage, p. 2, (Verlag Chemie, Weinheim 1990).
13. V. Gutmann and G. Resch, *Chim. Oggi* **10** (May 1992) 9.
14. L. L. Whyte, *Essay on Atomism - From Democritus to 1960,* p. 29, (Westeyan Univ. Press. Middletown, Conn. 1961).
15. W. Heisenberg, *Der Teil und das Ganze,* p.100, (Piper Verlag, Zürich, München, 1969).

CHAPTER 3

CHEMICAL BONDING

1. Introduction

Although the considerations of the preceding chapter seem self-evident, it has not been possible to apply their consequences to modern science so far. All efforts to construct a molecule from its isolated atoms have been based on *Newton's* approach and this requires the artificial introduction of certain "forces of attraction" between the bonded atoms. This approach has been useful in order to account for certain quantitative aspects but it cannot provide an understanding for the changes in qualities as a result of chemical changes.

2. Stoichiometry

It is well known that the representation of chemical changes by means of the chemical equation accounts for the changes in quantities, but it is not an equation with regard to the qualities. For example, the equation

$$C + O_2 = CO_2$$

means that 12 g carbon and 32 g oxygen give 44 g of carbon dioxide. The chemical symbols C and O_2 are separated on the left side of the equation and connected to each other on the right side to express different qualities. However, the quality of carbon dioxide cannot be understood by considering the qualities of the separated entities. Likewise knowledge of the properties of carbon dioxide gives no indication about the properties of its constituents.

It has been pointed out [1] that the different meanings of the Greek word "Analysis" are dissolution, elimination, redemption, liberation, departure, passing away, death. These express in different ways the irreversible loss of the functionalities of the parts as a result of their dismemberment from the system of higher complexity. The dismembered parts may be considered as "redempted" or "liberated" and this causes "death" of the superordinated system.

It has been clearly pointed out by *Kurnakov* [2] that the *law of constant and multiple proportions is the result of the application of the theory of numbers to chemistry* and that in this way the term "chemical composition" has been reduced to a mathematical term. When in modern chemical analysis the mass-spectrographic method is used, mass numbers are obtained (for ionized atoms), which are then translated into the symbols of the chemical elements. Thus, the concepts "equivalence" and "composition", as used in chemistry, correspond to those in the theory of numbers and the term "chemical compound" has actually been reduced to a mathematical term.

With regard to these quantitative relationships, the presentation of *Avogadro's* law should have provided strong support for *Dalton's* law. *Avogadro* had found in 1811 that equal volumes of different gases contain at the same temperature and at the same pressure the same number of molecules and that gases combine in simple proportions by volume. The number of molecules contained in one mole has been determined for the first time by *Loschmidt* in 1865 and is now called *Avogadro constant*. In order to understand *Dalton's* rejections of *Avogadro's* law we have to turn to the question, in which ways atoms are held together in molecules.

3. The Electrostatic Approach

It has been said that the connectivities of the atoms within a molecule are irretrievably lost by the dismemberment of the molecule into the atoms and hence artificial assumptions are made about binding forces.

Davy had realized connections between electrical and chemical phenomena and supposed that smallest particles may become oppositely charged as soon as they touched each other. As this was not expected to take place when two equal atoms touched each other, the formation of H_2 molecules or of Cl_2 molecules was considered as impossible. However their formation had been postulated by *Avogadro* as he had to replace the equation

$$H + Cl = HCl$$
$$\text{2 volumes} \quad \text{1 volume}$$

by

$$H_2 + Cl_2 = 2\,HCl$$
$$\text{2 volumes} \quad \text{2 volumes}$$

in order to meet the quantitative results. This was not accepted by *Dalton* and his contemporary scientists, because on ground of the electrostatic approach the union of two equal atoms was considered impossible. This thesis was proved correct by *Cannizzaro* as late as in 1858, two years after *Avogadro's* death.

Dalton's suggestion of the polarization of the atoms in the course of their interaction was further developed by *Berzelius*, who considered "compounds" as made up of electrically oppositly charged elements. This implied that neutral atoms have an "ability" to become charged and this has been proved correct by quantum mechanics, which considers charge transfer and polarization effects.

In his dualistic electrostatic theory, *Berzelius* had, however, not explicitly stated these abilities of the atoms, when he wrote: "we now believe that we know with certainty that bodies which are ready to be combined exhibit free opposite electricities." He was promptly critisized, as "free electricities" are not exhibited in the elements. Consequently his proposal of arranging the elements in the "polarity scale of the elements", advanced in 1818, was also rejected. This scale included 51 elements ranging from oxygen as the most electronegative element to potassium as the most electropositive element. Due to the rejection of this scale by the contemporary scientists of *Berzelius*, it had to be rediscovered more than hundred years later and is now well established as the electronegativity scale. Indeed the electronegativity is defined as an ability, as the "tendency of an atom" to attract electrons within a molecule and hence it is more related to quality than to quantity.

Further development of the electrostatic theory was, however, mainly due to quantitative studies. *Faraday* formulated the laws of electrolysis by establishing relationships between the quantity of electricity and the quantity of masses deposited at the electrode. The *Faraday-constant* (96 490 C) provided the bases for the determination of the elementary electric charge when the *Avogadro-constant* was known. The *Millikan*-experiment was originally designed and first performed by *Ehrenhaft* in 1910, who had, however, obtained a smaller value than $1.6 \cdot 10^{-19}$ C and was promptly considered to be wrong.

The electrostatic approach was further developed to the concept of the "heteropolar bond" between ions of opposite charge, held together by Coulombic forces. Later the crystal field theory was advanced and this was eventually applied to complex species after appropriate extension.

4. Covalency

The development of organic chemistry in the 19th century is characterized by qualitative advances. The credit for the formulation of valency is given by the historians of chemistry to *Frankland*, who had the idea, in 1852, that an element can combine only with a certain limited number of atoms [3]. An enormous advance was made by *LeBell* and *van't Hoff* in 1874 by their suggestion of the tetrahedral arrangement of the atoms around a carbon atom [4]. This was based on qualitative considerations. This thesis was strongly opposed by *Kolbe* [5], who called this work "an essay full of fantastical speculations in which the investigation of the spatial arrangement of atoms is undertaken with a boldness which surprises the serious scientist and is not far removed from a belief in sorcery and necromancy".

After the discovery of the electron, *Thomson* proposed the hypothesis that electrons take part in chemical bonds. This idea was further advanced by *Lewis* [6] in 1917, who suggested that the valence-dash may be interpreted by the sharing of a pair of valence electrons. He considered the atomic shells as mutually interpenetrable and arranged the electrons at the corners of a cube. He suggested that the bonded atoms touch each other with the edges in case of a single bond and with faces in case of a double bond. In this way he interpreted each valence dash by a pair of electrons. In 1923 he proposed the electronic theory of acids and bases, defining an acid as an electron-pair acceptor and a base as an electron-pair donor [7]. He was, however, ridiculed for the "mystic tendency of the electrons to arrange themselves pairwise in certain directions of the space". Although in his simplified model the electrons were considered at more or less fixed positions, his interpretation is now considered to be basically correct [8].

5. Advances and Limitations of Quantum Chemistry

In this context it is impossible to pay adequate attention to the development of numerical quantum chemistry. This has been developed as a consequence of the fact that the *Schrödinger*-equation cannot be solved for multi-body systems.

The orbital Ψ is strictly a mathematical term, which is not subject to any physical interpretation. Ψ^2 has been associated with the picture of an "electron cloud". Although it is impossible to define border lines for it, an orbital is commonly considered as a region in space around the nucleus that can be occupied by no more than one pair of electrons. An orbital is considered as a region within which there is a 90 to 95 % probability to find the electron. The shape of an atomic orbital is depicted in a three-dimensional figure confined by artificially constructed border lines. This pictorial description by means of the graphical representation of orbitals and the hybridization theory leads to the illusion of providing an understanding. It is, however, unsuitable for this purpose, because the orbitals penetrate each other and it is virtually impossible to provide experimental evidence for the correctness of the graphical representation.

It has also been clearly expressed in the Copenhagen interpretation that quantum mechanics deals only with certain aspects of the microcosms and not with the things as they are found in nature. Reality has been restricted to those aspects which have already been perceived according to the agreement paraphrased as the "principle of unreality of unobserved microevents" [9]. Accordingly "reality as it is" is considered a "metaphysical speculation outside physics"[1].

The difficulties in quantum mechanics to provide either a qualitative understanding or to lead to precise mathematical results, have been further increased by the so-called *Einstein - Podolsky - Rosen* correlations [10]. In these it has been shown that quantum theory is not a complete theory, as it does not even attempt to describe certain important aspects of reality, namely the fact that all partial systems are interrelated and that the existence of the objects, which can be isolated, cannot be maintained. *Bohm* and *Aharanov* [11] advanced the following argument:

Consider a pair of spin one half particles, A and B, formed in the singlet spin state and moving freely in opposite directions. Measurements can be made - say by *Stern-Gerlach* magnets - of each of the components with opposite spin. Since we can predict the result of a measurement on B after having measured the properties of A, it follows that the result of any such measurement must be predetermined. According to *Primas* [12] the *Einstein-Podolsky-Rosen* correlations have also been proved experimentally.

It has been argued that particle A seems to know the state of particle B. This connection should even enable the investigator to determine the state B by influencing the state A. *Bell* [13] emphasized that there must be a mechanism

---

[1] In this statement the fact is ignored that abstraction itself is subject to metaphysical criteria, as these are determined by the investigator. It is true that abstractions are necessary, but they are justified only as long as they are used as tools in order to gain knowledge; it is, however, unrealistic to consider the result of an abstraction as scientific knowledge about the things as they are.

whereby the setting of one measuring device can influence the reading of another instrument, however remote, and that this fact is incompatible with the statistical predictions of quantum mechanics.

This state of affairs prompted *Schrödinger* [14] to the following remark: "It is rather discomforting that the theory should allow a system to be steered and piloted into one or the other type of state at the experimenter's mercy in spite of his having no access to it".

This means that *quantum systems which have been interrelated in the past, are in all future in a holistic way correlated to each other* [12]. It is therefore impossible to put "independent" parts together, as is attempted in all approaches in theoretical chemistry.

Despite this convincing result, the *Cartesian* approach is used in quantum chemical calculations by introduction of parts assumed to be independent. In all these attempts the *Einstein-Podolsky-Rosen* correlations are eliminated [12] and even classical forces introduced, which result exclusively from theoretical considerations and which can never be observed or measured. Nevertheless such interaction forces have been given ontological character, and their characterization by names, such as *van der Waals* forces, *Coulomb* forces, cohesion forces etc. has contributed to a mistaken belief in their real existence. Actually, they have been "invented" to fit mathematical requirements.

Not even quantum theory or a kind of "super-theory" will be able to completely describe reality. The Copenhagen interpretation has actually capitulated before reality, because it allows the description of the world only by means of theories, which can never reach reality.

As quantum mechanics describes the whole universe as one continuum, fluctuating in space and time, it takes into account the interrelationships and the interdependencies between all of the observable discontinuities. Quantum mechanics is therefore concerned only with one of the *Aristotelian* categories, namely that of the *relations*.[2] Because quantum theory deals with the category of relations within the continuum and because physics is concerned with relations, it does provide the *most universal and most useful theory that has ever been presented for an understanding of the relationships*. These are important and require that the observable discontinuities "behave" in certain characteristic ways (see Chapter 16).

References

1.   V. Gutmann, *Fresenius Z. analyt. Chem.* **337** (1990) 166.
2.   N. S. Kurnakov, *Z. anorg. allg. Chem.* **88** (1914) 143.

---

[2] Quantum mechanics is not concerned with the *Aristotelian* categories of substance, quality and quantity (see Chapter 16). The category of substance has never been considered in modern science. For the understanding of quality the orbital representations are not suitable. With regard to quantity, all efforts to perform precise calculations have not been successful. Because of the motions in the continuum the objects cannot be precisely defined and hence all hopes to do better by using more efficient computers are unjustified for this fundamental reason.

3. R. Mierzecki, *Historical Development of Chemical Concepts*, p. 175, (Polish Soc. Publ. Warszawa, 1991).
4. H. S. van Klooster, *J. Chem. Educ.* **29** (1952) 367.
5. E. Cohen, *Jakobus Henricus van't Hoff, sein Leben und Wirken*, (Akad. Verlagsges. Leipzig, 1912).
6. G. N. Lewis, *Valence and Structure of Atoms and Molecules* (The Chemical Catalogue Co. New York 1923).
7. G. N. Lewis, *J. Franklin Inst.* **226** (1923) 293.
8. W. F. Luder and S. Zuffanti, *The Electronic Theory of Acids and Bases* (Wiley, New York, 1946).
9. H. Mehlberg, in: *Quantum Theory and Reality*, Ed. M. Bunge, p. 48, (Springer, Berlin, Heidelberg, New York 1967).
10. A. Einstein, P. Podolsky and N. Rosen, *Phys. Rev.* **47** (1935) 777.
11. D. Bohm and Y. Aharonov, *Phys. Rev.* **108** (1952) 166.
12. H. Primas, *Chemistry, Quantum Mechanics and Reductionism*, 2nd ed., p.142, (Springer Berlin, Heidelberg, New York, 1983).
13. J. S. Bell, *Physics* I (1964)195.
14. E. Schrödinger, *Proc. Cambridge Phil. Soc.* **31** (1935) 556.

CHAPTER 4

INTERACTIONS BETWEEN MOLECULES

1. General

It has been emphasized that the continuum conditions which are neglected in considerations of the molecular concept, are, at least in principle, considered by quantum mechanics. Because these connections and continuous relationships are not directly observable, conclusions have to be drawn from observable changes of the discontinuous parts within the continuum. *The regularities, which are found, require that the parts "behave" in certain characteristic ways.*

In order to describe their behaviour, the term "interactions" is used. The interactions are attributed to individual properties of atoms or molecules and one is no longer aware of the requirement of the presence of an unobservable continuum for the interactions to take place.

In order to observe regularities, the observed system must have an "ability" to react in highly specific ways towards changes in the environment, whereby *their main characteristics and functionalities are maintained* [1]. This requires that each observed object must be in a position to minimise the action of forces from outside by developing appropriate forces against them or by taking measures in order to "escape" them.

Such abilities are known for macroscopic systems and they are necessary for the fulfilment of the most fundamental physical laws, although these abilities are not explicitly stated. The laws of conservation of mass and energy, as well as the law of inertia, require that forces are put into operation so as to respond to the forces by which the system is changed. The theorem of *LeChatelier-Braun* states: "Wherever a constraint is placed on a system in equilibrium, the latter is altered in such a way as to tend to minimise the effect of the constraint". The development of forces against the actions of external forces is clearly expressed in *Lenz's* law in electromagnetism. It states that an induced electric current is made to flow in such direction that the current opposes the change that produced it. Even the energy principle requires developing forces as to reach and to maintain the state of minimum energy.

Because quantum mechanics is mainly concerned with the mathematical formulation of the relations, efforts have been made by chemists to express the relations in terms more familiar to them. The concern of chemists with changes in electron densities led to the description of molecular interactions by considering one of the reacting molecules to act as the donor of electronic charge and the other one as the acceptor of electronic charge.

Chemists are used to distinguish redox-reactions from acid-base reactions (in the *Lewis* sense). On a molecular level redox-reactions are idealistically described by complete transfer of one or of more electrons between the reducing and the oxidizing agent. The former is considered as an electron donor (ED) and the latter as an electron acceptor (EA). Acid-base interactions are interpreted on the molecular level by sharing an electron pair that is provided by the electron pair donor (EPD), the Lewis base and accepted by the electron pair acceptor (EPA), the Lewis acid.

2. Chemical Functionality

The concept of chemical functionality [2,3] is based on these idealistic descriptions and it takes into account that the actual changes in net-charges due to an electron transfer reaction (redox reaction) are smaller than the charges of the electrons transferred. It takes also into account that by sharing an electron pair changes in net-charges are bound to take place. It is therefore assumed that in the course of a redox reaction within the developing species an acid-base interaction is induced to compensate in part the effect of the electron transfer.

For example, in the course of the redox reaction

$$[Fe(OH_2)_6]^{2+} \rightleftharpoons [Fe(OH_2)_6]^{3+} + e^-$$
$$\text{ED} \qquad\qquad \text{EPA}$$

the Fe(II)-species acts as an ED, but its loss in electron charge by conversion into Fe(III) is partly compensated by the increased EPA-functions of Fe(III). This leads to a strengthening and shortening of the Fe - OH_2 bonds. By the stronger hydration of Fe(III) the positive net-charge at the iron nucleus is decreased and partly spread over all of the bonded water molecules with slight increases of positive net-charges at each of the hydrogen atoms of the bonded water molecules (see Chapter 14).

On the other hand, an EPD-EPA interaction is bound to lead to changes in net-charges although the oxidation numbers are not altered. This is well illustrated for the ionization of a covalent substrate in an appropriate solvent [4]. For example, the dissolution of $LiCH_3$ in ether involves EPD-interactions of the ether molecules with the Li-site of the solute, which acts as an EPA towards the ether molecules. As the positive charge at the Li-site is decreased, an electron shift is initiated from the Li-site, which acts as an ED, to the CH_3 group, which acts as an EA to produce the charged ions with increase in positive net-charge at a former and with increase in negative net-charge at the latter:

$$n\,R_2O + Li\text{-}CH_3 \rightleftharpoons [Li(OR_2)_n]^+ + CH_3^-$$
$$\text{EPD} \quad\text{EPA}$$
$$\text{ED} \quad\text{EA}$$

According to this approach, the ionization of hydrogen chloride in water is described by

(i) the EPA function of the hydrogen atom of the HCl molecule and the EPD functions of the coordinating water molecules and

(ii) the EPD function of the chlorine atom of the HCl molecule towards water molecules acting as EPA. Hydration at the hydrogen atom leads to a decrease in positive net-charge at the hydrogen and hydration at the chlorine atom to a decrease in negative net-charge at the chlorine atom. These effects are overcompensated by the development of the ED function at the hydrogen atom and of the EA function at the chorine atom [4], i.e. the shift of electronic charge from the hydrogen atom towards the chlorine atom of the hydrogen chloride molecule with subsequent formation of the hydrated ions:

$$\text{n (H}_2\text{O)} \rightarrow \text{H - Cl} \rightarrow \text{(HOH)}_m \rightleftharpoons \text{[H (OH}_2)_n]^+ + \text{[Cl (HOH)}_m]^-$$
$$\text{EPD} \qquad \text{EPA EPD} \qquad \text{EPA}$$
$$\text{ED} \quad \text{EA}$$

The concept of chemical functionality found little acceptance by chemists, possibly because no quantitative correlations between donor functions and acceptor functions on the one side and changes in net-charges on the other side can be established for the simple reason that it is impossible to define and to measure the values for the net-charges of atoms within molecules.

3. Variations in Bond Lengths

Introduction
The relations of the changes in electron densities for the changes in structural parameters, namely in bond lengths and in bond angles have been formulated by rules. These are called the bond length variation rules [6] and sometimes *Gutmann-rules* [7]. *They are based on qualitative observations and on quantitative measurements, in agreement with quantum chemical requirements and results and completely independent of any model assumption about chemical bonding.*

According to quantum mechanics, charge transfer is involved in any molecular interaction. This means that one of the reacting molecules acts as a donor of electronic charge and the other one as an acceptor of electronic charge. Any such changes are reflected not only in changes in electron densities, but also in changes in the positions of the atoms, measurable by changes in bond lengths and changes in bond angles. Because the atoms provide observable discontinuities within the charge density pattern, it has been suggested to consider in the first place the changes in bond length and bond angles, which are related to appropriate changes in electron densities.

The First Bond Length Variation Rule and the Pileup and the Spillover Effects.
The first bond length variation rule states that an intermolecular interaction leads to lengthening of the intramolecular bonds which are adjacent to the site of the intermolecular interaction [6]. The shorter the intermolecular distance D → A, the greater is the lengthening of the adjacent intramolecular bonds both in the donor and acceptor component:

An increase in intranuclear distance is indicated by a full bent arrow pointing in the direction of the charge transfer between the two considered atoms.

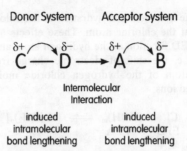

Inverse relationships between intermolecular and intramolecular bond length are well demonstrated by the results of X-ray analyses of different crystalline hydrates [8] and shown in Fig. 1.

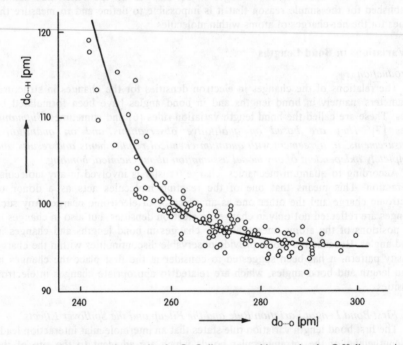

Fig. 1 Relationship between intermolecular O···O distances and intramolecular O-H distances in crystalline hydrates [8].

The changes are well demonstrated for the interaction between NH_3 and BF_3:

Donor System   Acceptor System

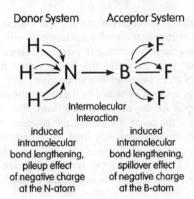

Intermolecular
Interaction

induced   induced
intramolecular   intramolecular
bond lengthening,   bond lengthening,
pileup effect   spillover effect
of negative charge   of negative charge
at the N-atom   at the B-atom

Table 1. Changes in intramolecular B-F bond length as related to the intermolecular N-B distances

Compound	Intermolecular N - B distance [pm]	Intramolecular B - F distance [pm]
BF_3	-	130
$CH_3CN - BF_3$	163	133
$H_3N - BF_3$	160	138
$(CH_3)_3N - BF_3$	158	139

Bond lengthening is associated with an increase in bond polarity [9] and hence with increasing fractional charges of the bonded atoms. Experimental evidence, as well as results of quantum-chemical calculations reveals that both the negative net charge of the donor atom and the positive net charge of the acceptor atom are usually increased. The loss of negative charge at the donor atom due to interaction with the acceptor atom is overcompensated by the induced attraction of negative charge from the other atoms of the donor unit. This is called the *pileup effect* of negative charge at the donor atom [6,10].

On the other hand, charge transfer to the acceptor atom is overcompensated in the latter by an induced flow of negative charge from the acceptor atom towards the peripheral atoms. This leads to an increase of positive charge at the acceptor atom and of negative charge of the peripheral atoms and is referred to as the *spillover effect* of negative charge at the acceptor atom [6,10].

These effects are essential requirements for the course of the ionization of a covalent substrate. The representation of a nucleophilic substitution by an amine in organic chemistry by assigning an increase in positive charge to the nitrogen atom of the amine is not correct. It is true that the whole amine molecule becomes more positive, but the nitrogen atom becomes more negative.

These considerations are not confined to Lewis acid-base interactions, but may be applied to other types of molecular interactions. For example, in the iodine molecule in the gas phase, the mean intramolecular distance is 268 pm and the mean internuclear distances are variable and very long. However, in the crystal, because of the relatively short intermolecular distance of 345 pm, the mean intramolecular distance is increased to 272 pm. Even more considerable are the differences in M-X distances in alkali halides in the gas phase and in the crystalline state[1], as seen from Table 2.

Table 2. Alkali metal - halogen bond distances d (in pm) in the diatomic molecules in the gas phase and in the crystal lattice in the solid state.

Halide	Gas	Crystal	Δd
LiBr	217	275	26.5 %
NaBr	250	298	19.1 %
KBr	282	329	16.8 %
RbBr	294	343	16.8 %
KCl	267	315	18.0 %
RbCl	279	329	17.9 %
LiI	239	303	26.5 %
NaI	271	323	19.1 %
KI	305	353	16.8 %
RbI	318	366	15.1 %

Under pressure, the intermolecular distance in iodine crystals is further decreased and consequently the intramolecular distance increased, until all I-I distances are equally long (287 pm) [6]. The so-called *pressure distance paradox* expresses the fact that increase in pressure causes lengthening of the chemical bonds in crystals, although at the same time the density is increased and hence the volume available for the atoms is decreased.

The solution to this paradox is provided by the bond length variation rules [11]. The "intermolecular" distances (e.g. in silicon dioxide the O-O and Si-Si distances) are commonly not considered as chemical bonds. Under the action of

---

[1] This shows that the individual character of molecules is more pronounced in the gas phase than in either of the condensed states.

pressure they are shortened to a greater extent because of the low electron densities between them, and the intramolecular distances, denoted as chemical bonds (e.g. in SiO_2 the Si-O distances) are lengthened accordingly. The transformation of coesite (SiO_2) into stishovite (SiO_2), is accomplished at 100 kbar and 1200°C. As the density is increased under these conditions from 2.93 to 4.28 g.cm^{-3}, the intermolecular distances are shortened, namely the Si-Si distances from 301 pm in coesite to 267 pm in stishovite and the O - O distances from 263 to 251 pm, respectively. Consequently, the Si-O bond lengths are increased from 161 pm in coesite to 178 pm in stishovite.

A further example of this paradox is the transition of graphite into diamond. The C-C distances in diamond are greater than those within the graphite layers. The increase in density in going from graphite to diamond is associated with a considerable shortening of the interlayer distances initially present in graphite and hence with a slight lengthening of the C-C distances, formally within the layers.

Because the molecules at a surface of a condensed phase (solid or liquid) have smaller coordination numbers than within the bulk of the phase, a bond shortening at a surface is expected [12]. *Lennard-Jones* [13] predicted in 1927 that the lattice parameters near and perpendicular to the surface should be smaller than those within the crystal lattice. Due to the lack of experimental techniques, no experimental evidence has been found for a long time, but 50 years later, the low-energy electron diffraction method allowed its confirmation. The results revealed, for example, that the mean internuclear distances in silver (111) surface planes are about 6 % smaller than those within the crystal.

The first bond length variation rule may also be formulated in terms of coordination numbers. As the coordination number is increased, the length of the bonds originating from the coordination centre is increased and this is an extension of the *Goldschmidt* rule [14] formulated for solid state systems.

The Second Bond Length Variation Rule

The second bond length variation rule concerns the description of changes in bond lengths in all other regions within the newly formed molecular system. It states that *bond lengthening occurs when negative charge is shifted from a more electropositive towards a more electronegative atom* (in agreement with the first rule) and that *bond shortening takes place when negative charge is shifted from a more electronegative towards a more electropositive atom*. Bond polarity and bond length are increased in the first case, but decreased in the second. A broken line arrow is used to indicate a charge transfer associated with shortening of a bond, i.e. its strengthening.

Alternating bond lengthening and bond shortening is found in those molecules in which atoms of higher and lower electronegativities are arranged in alternating sequences. This may be illustrated in Fig.2 by the interaction of antimony pentachloride with tetrachloroethylenecarbonate [15]. According to the first bond length variation rule, the intramolecular bonds, which are adjacent to the intermolecular O→Sb bond, namely the C=O bonds and the Sb-Cl bonds, are lengthened (electron shift from the more electropositive towards the more electronegative atom) whereas the neighbouring O-C bonds are shortened (electron shift from the more electronegative to the more electropositive atom). The adjacent C-O bonds are lengthened and, in the peripheral regions of the donor component tetrachloroethylenecarbonate, both the C-C bonds and the C-Cl bonds are shortened as negative charges from the chlorine atoms are shifted towards the peripheral regions of the acceptor molecule SbCl$_5$, thereby decreasing the negative charges of the terminating chlorine atoms bonded to the C-atoms.

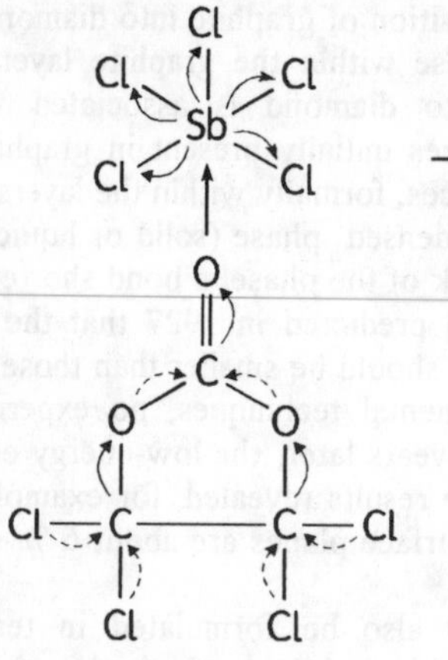

Bond	Bond lengths in the isolated molecules [pm]	Bond lengths in the adduct [pm]
Sb-Cl	231 - 243	235 - 247
C=O	115	122
O-C	133	125
C-O	140	144
C-C	153	143
C-Cl	176	174

Fig. 2. Bond length in the free molecules (SbCl$_5$ and tetrachloroethylenecarbonate) and in the adduct.

Due to the intermolecular donor-acceptor interaction the whole system becomes more polarized and more reactive [16]. The increase in negative charge in the outer regions of the acceptor molecule has been demonstrated by the changes in the ^{19}F-NMR chemical shift in CF$_3$I dissolved in donor solvents [17]. The donor solvent D attacks the iodine atom, from which negative charge is passed on through the carbon atom towards the fluorine atoms which are increased in negative charge in relation of the solvent donor number.

4. Variations in Bond Angles

In agreement with the *Gillespie-Nyholm* rules [19] coordination of the donor molecule leads to an increase of the adjacent bond angles in the donor system, whereas coordination of the acceptor molecule leads to the formation of a new bond and hence to a decrease of the existing bond angles in the acceptor molecule. This leads to the following rule: As the intermolecular interaction increases the *intramolecular angles of the donor component* (measured opposite to the attack of the acceptor unit) *are increased and the intramolecular angles of the acceptor component* (measured opposite to the attack of the donor unit) *are decreased.*

Donor System   Acceptor System
Bond angle increased   Bond angle decreased

In other words, in the course of the donor-acceptor interaction, the atoms of the donor system which are adjacent to the interaction site bend towards the acceptor system, whereas in the acceptor system the adjacent atoms bend away from the approaching donor system [20]. The previously mentioned nitrogen donor-boron trifluoride system may be taken as an example for the changes in the acceptor component (Table 3).

Table 3. Changes in bond length in the acceptor component BF_3 with increasing intermolecular bond strength.

Compound	F - B - F bond angle
BF_3	120°
$CH_3CN - BF_3$	114°
$H_3N - BF_3$	111°
$(CH_3)_3N - BF_3$	107°

In agreement with these rules, the Cl-Sn-C angle in trimethyltinchloride is decreased as the donor number of the donor D is increased. For HMPA, the strongest donor solvent investigated, an angle of 94.3° is found, which is close to that of 95.7° reported for $(CH_3)_3SnCl.HMPA$ in the solid state from X-ray diffraction data. A clear relationship is found between the mean values of the C-Sn-Cl angles and the solvent donor number showing that the bipyramidal configuration is more closely approached as the donor strength of the solvent is increased.

Fig. 3 shows the relations between the intramolecular Sn-Cl bond length and the Cl-Sn-C bond angle as obtained by X-ray diffraction data.

Changes in the charge distributions and geometries for a series of adduct molecules have been calculated by means of quantum chemical methods and these are in good agreement with the results of spectroscopic measurements. There is a continuous set of molecular geometries between tetrahedral and trigonal bipyramidal arrangement, the latter being approached more closely as the solvent donor number is increased.

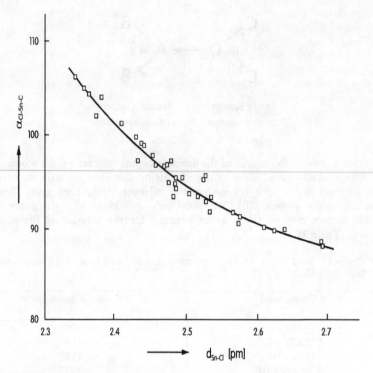

Fig. 3. Sn-Cl distances versus Cl-Sn-C angles in solid adducts of trimethyltinchloride .

5. Effects at Terminal Positions

Studies of the bond energies in crystals and in solutions have revealed *that the electronic changes are well-pronounced in the terminal regions of the system* under consideration. For example, *Singh* and *Tedder* [25] found an increase in rate of chlorination of carboxylic acids after coordination of acceptor molecules at the C=O group which is greatest at the C-H group terminating the chain.

The results obtained for the interactions of ammonia with boranes carrying linear alkyl substituents of varying chain length confirm the great influence of the charge transfer between donor and acceptor atoms at the C_1 atom *and* at the terminating C atom [26]. Alternating changes in the charges are least pronounced in the middle of the carbon chain. The greatest changes are found at the terminating atoms which become more negative at the acceptor component (the boranes) and more positive at the donor component (the amines), in agreement with the bond length variation rule and the spillover and pileup effects, respectively.

A convincing example for the *pronounced effects at the end of the molecular chain* is [27], provided by haemoglobin, where two protons are released from the ends of the protein chain separated by 146 amino acid residues from the iron atom of the haeme groups as soon as oxygen has been coordinated to the latter [28,29].

Haemoglobin is contained in the red blood cells, the erythrocytes of vertebrates. Its purpose is to carry oxygen from the lungs to the tissues, as well as to help to transport carbon dioxide, the waste product of metabolism, back to the lungs, where it is excreted. Because the red blood cell does not contain a nucleus, it lacks static boundary conditions not only inside the cell, but also within the membrane.

A haemoglobin molecule (molecular weight 64450) is made up of four polypeptide chains (linear sequences of amino-acid residues), namely two alpha chains of 141 amino-acid residues each and two beta chains of 146 residues each. The alpha and beta chains have different sequences of amino-acids, but fold up to form similar three-dimensional structures [28,29]. Each chain harbours one haeme, which gives blood its red colour, and which is the active centre, the binding site for oxygen. The haeme is a flat ring, called a porphyrin "with an iron atom sticking out of its centre like a jewel" [28]. Each haeme group is connected to the protein by means of a histidine group.

By coordination of molecular oxygen to an iron atom, the bond in the oxygen molecule is slightly lengthened and the iron atom pushed into the plane of the porphyrin ring and these changes have enormous consequences.

1. Oxygen coordination at one of the haeme groups leads to appropriately increased reactivities of the other three haeme groups towards oxygen. This has been expressed by *Perutz* [28] as follows: "Oxygen-free molecules (desoxyhaemoglobin) are reluctant to take up the first oxygen molecule, but their appetite for oxygen grows with the eating. Conversely, the loss of oxygen by some of the haemes lowers the oxygen affinity of the remainder. The distribution of oxygen among the haemoglobin molecules in a solution therefore follows the biblical parable of the rich and the poor: "For unto everyone that hath shall be given, and he shall have abundance; but from him that hath not, shall be taken away even that which he hath".

This phenomenon shows that there are excellent communications between the haemes in each molecule to which physiologists refer to as haeme - haeme interactions.

2. Oxygen coordination is followed by an enormous conformational change of the polypeptide chains so that the oxygen molecules were compared by *Perutz* [28] with "a flee that makes an elephant jump". In the course of the conformational change the strong bonds between $alpha_1$ and $alpha_2$ haemes, as well as between $beta_1$ and $beta_2$ haemes are loosened, whereas the weak bonds between $alpha_1$ and $beta_1$, as well as those between $alpha_2$ and $beta_2$ haemes are strengthened.

When oxygen is released at the tissues, all of the effects are reversed, so that *Perutz* [28] compared the conformational changes with "a rock and roll movement" of the dimer $alpha_1$-$beta_1$ with respect to the dimer $alpha_2$-$beta_2$.

3. Based on structural investigations the coordinated oxygen molecules appear to be rotating within the pockets and these rotations must be in accord with the movements of the polypeptide chains forming the pockets. The haeme pockets are narrower in the T-structure (in the absence of oxygen) than in the R-structure (after oxygen coordination). The former is under greater tension than the latter.

4. The transformation of haemoglobin into oxyhaemoglobin is accompanied by the intramolecular transformation from the paramagnetic high spin state into the diamagnetic low spin state, which is energetically more favourable. In the course of these transitions the system has an optimal flexibility and adaptability towards the environment while maintaining its integrity over the temperature range between +34° and +42°C [27].

5. For every four molecules of oxygen taken up at the lung, two protons are released into the cell plasm from the terminal positions of the protein chains, separated from the iron nuclei by more than 140 amino acid residues. In this way carbon dioxide is set free at the lungs from hydrogen carbonate ions dissolved in the plasm:

$$\text{haemoglobin} + O_2 \rightarrow \text{oxyhaemoglobin} + 2\,H^+$$

$$2\,H^+ + 2\,[HCO_3]^- \rightarrow 2\,CO_2 + 2\,H_2O$$

In turn, for every four oxygen molecules released at the tissues, two protons are taken up from the cell plasm in order to allow carbon dioxide to pass into solution in the cell plasm in form of hydrogen carbonate ions:

$$\text{oxyhaemoglobin} + 2\,H^+ \rightarrow \text{haemoglobin} + O_2$$

$$2\,CO_2 + 2\,H_2O \rightarrow 2\,H^+ + 2\,[HCO_3]^-$$

The latter process is catalysed by the enzyme carbonic anhydrase which is present in the red blood cells. In this enzyme a zinc ion is situated at the active site (bound to three histidine groups) which carries water as its fourth ligand that is displaced in the process of catalysis [30].

The interactions between haemoglobin and oxygen lead, therefore, not only to dramatic changes within the haemoglobin molecule but also to changes in its natural environment, the cell plasm of the erythrocyte. These changes are required for the dissolution of carbon dioxide at the tissues, for its transportation and for its release as carbon dioxide at the lungs.

The dramatic changes in the peripheric regions of the haemoglobin molecule caused by coordination and release of oxygen respectively provide convincing evidence not only for the far-reaching effects of intermolecular interactions and their effects at the terminal positions, but also for the mutual interactions with the aqueous phase.

6. "Intelligent Behaviour"

The interdependencies and specific interactions between haemoglobin and oxygen at the one side and between haemoglobin and its surrounding cell plasm on the other side suggest an "intelligent behaviour" on a molecular level. *Eigen* and *Winkler* [31] refer to molecules which respond in highly specific ways to changes in the environment as *"intelligent molecules"*. Under this title a paper has appeared recently [1] in order to express properties which cannot be measured, namely the response of molecules to slight changes in environment as well as the specific actions of molecules towards the environment without losing their integral configurations and functionalities.

However, intelligence cannot refer to the molecules themselves[2] but rather to the source of their organizing ability (be it chance or the creator). We shall come back to these points in Chapter 21.

References

1. V. Gutmann and G. Resch, *Chem. Int.* **10** (1988) 5.
2. V. Gutmann, *Chemische Funktionslehre* (Springer Verlag Wien, 1971).
3. V. Gutmann, *Monatsh. Chem.* **102** (1971) 1.
4. V. Gutmann, *Angew. Chem.* **82** (1970) 858; *Int. Ed.* **9** (1970) 843.
5. V. Gutmann and U. Mayer, *Monatsh. Chem.* **100** (1969) 2048.
6. V. Gutmann, *The Donor-Acceptor Approach to Molecular Interactions*, (Plenum Press, New York, 1977).
7. W. B. Jensen, *The Lewis Acid - Base Concept, an Overview* (Wiley and Sons, New York 1980).
8. J. O. Lundgren and I. Olovsson, *in Hydrogen Bond, Recent Developments in Theory and Experiment*, eds. P. Schuster et al. (North Holland Publ. Co. Amsterdam, 1976).
9. I. Lindqvist, *Inorganic Adduct Molecules of Oxo-Compounds*, (Springer, Berlin, Göttingen, Heidelberg, 1963).
10. V. Gutmann, *Rev. Roum. Chim.* **22** (1977) 619.
11. V. Gutmann and H. Mayer, *Structure and Bonding* **31** (1976) 328.
12. V. Gutmann and H. Mayer, *Rev. Inorg. Chem.* **1** (1979) 51.
13. E. J. Lennard - Jones, *Proc. Roy. Soc. London* **A 121** (1928) 247.
14. V. M. Goldschmid and A. Muir, *Geochemistry* (Clarendon Press, Oxford 1958).
15. H. Kietaibl, H. Völlenkle and A. Wittmann, *Monatsh. Chem.* **103** (1972) 793.
16. V. Gutmann and G. Resch, *Z. Chem.* **19** (1976) 406.
17. M. P. Spaziante and V. Gutmann, *Inorg. Chim. Acta,* **5** (1971) 273.
18. D. Hankins, J. W. Moskowitz and F. E. Stillinger, *J. chem. Phys.* **53** (1970)

---

[2] This has dramatic consequences for our own intelligence, which in its structure is bound to follow the structure as we observe and perceive it in nature.

4544.
19. R. J. Gillespie and R. S. Nyholm, *Quart. Revs.* **11** (1957) 339.
20. W. Linert and V. Gutmann, *Coord. Chem. Revs.* **117** (1992) 859.
21. W. Linert and V. Gutmann, *Rev. Chim. Min.* **20** (1983) 516.
22. W. Linert, A. Sotriffer and V. Gutmann, *J. Coord. Chem.* **22** (1990) 21.
23. W. Linert, V. Gutmann and A. Sotriffer, *Vibr. Spectrosc.* **1** (1990) 199.
24. W. Beck, A. Melnikoff and R. Stahl, *Chem. Ber.* **99** (1966) 3721.
25. H. Singh and J. M. Tedder, *J. chem. Soc.* **B** (1966) 603.
26. W. Linert, V. Gutmann and P. G. Perkins, *Inorg. Chim. Acta,* **69** (1983) 61.
27. G. Resch and V. Gutmann, *Scientific Foundations of Homeopathy* (Barthel Publ. Germany, 1987).
28. M. F. Perutz, *Scientific American,* (Nov. 1964) .
29. M. F. Perutz, *Nature* **228** (1970) 726.
30. A. Liljas et al., *Nature, New Biol.* **235** (1972) 131.
31. M. Eigen and R. Winkler, *Das Spiel,* p. 43 (Piper Verlag, München, Zürich, 1965).

CHAPTER 5

THE LIQUID STATE

1. Macroscopic Properties

Although liquid water plays an enormous role on earth and although most chemical reactions take place in its liquid solutions, all attempts in science to provide a proper understanding of the liquid state have not been very successful.

The liquid state of a material may be reached either by heating the solid or by cooling the gas. Many properties show also that *the liquid state is somewhere between the solid and the gaseous state*. For example the values for entropy and heat capacity of the liquid are greater than those of the solid, but much smaller than those of the gas. They are substantially increased by evaporating the liquid. Likewise, the values for density and compressibility are smaller in the solid and considerably greater in the gas than in the liquid.

Both a solid and a liquid have a definite volume and they are characterized by phase boundaries. On the other hand a gas has no definite volume and accepts the phase boundaries as they are provided by those of the former. Many solid phases are known, but only a limited number of liquid phases can exist in contact with each other such as water and ethyl ether or mercury and most other liquids.

The phase boundaries separate the two adjacent phases and connect them at the same time with each other, so that exchange of matter and energy is taking place in either direction. This means that there are no absolutely rigid clear-cut borderlines in nature and that a liquid touching a solid must have certain aspects of the latter and vice versa. Likewise the liquid surface adjacent to another liquid or to a gas must have certain aspects of both liquids and of the liquid and of the gas respectively.

It is also well-known that molecules of the liquid evaporate at a liquid-gas interface where molecules of the gas phase condense on the liquid. These interactions are continuously taking place between the different phases.

It is not possible to superheat a solid material to a temperature where the liquid is the thermodynamically stable phase. However, a liquid may be superheated to a temperature, where the gas phase is thermodynamically stable, provided that the liquid is clean and heated in a smooth vessel. When the superheated liquid changes to the gas, it does so with a bumping noise. In contrast to a gas a liquid may also be supercooled to below its freezing temperature and this is quite easy to do under carefully controlled conditions (see Chapter 6).

All of these properties imply that the conservative static aspects are best developed in a solid material, less in a liquid and hardly found at all in a gas. Thus, there is an increase in the dynamic (dissipative) aspects in going from the solid to the liquid and further to the gaseous state.

A system in the gaseous state appears to approach more closely the conditions as they seem characteristic for the non-differentiable continuum (Chapter 2). On the other hand, a system in the solid state appears well-contrasted from the non-differentiable continuum. It may be characterized by the domination of the observable conservative aspects of order, whereas in a gas the dynamic aspects of order are dominating.

In the liquid state the dynamic aspects of order are less pronounced than in the gaseous state and the static aspects of order less developed than in the solid state. In the solid state the hardly pronounced dynamic aspects of order cannot be directly observed, but they may be indirectly inferred and they are described as "displacement reactions". In the gaseous state the static aspects of order are not directly accessible. So it seems as though nature had placed the most significant aspects in such ways as to render them hardly accessible to man.

2. Molecular and Structural Properties

All of these facts are reflected in the properties on the molecular level. The differences of the interactions between the molecules in a given solid material before and after melting to a liquid are so small that - according to statistical thermodynamics - melting of a solid or freezing of a liquid would not be expected. At this point, the solid and the liquid phases are in equilibrium with nearly the same number of building units per unit volume in each of the phases and hence the mean energies per part are nearly the same. For these reasons statistical thermodynamics cannot predict or understand the differences in the solid and in the liquid state, and hence only one of the condensed phases would be expected to exist.

Although the X-ray spectra show differences for crystalline materials and for liquids, such as water, they reveal also certain similarities at least near the freezing point of the liquid, so that in this way at least limited access is provided to the structural features of a liquid. In the liquid state the molecules are known to move much more freely than in the solid state, whereas in a gas the molecules appear to move about nearly free with respect to each other in a large volume.

In the absence of observable structural boundary conditions in a gas, no restrictions can be observed for the movements of the molecules. This indicates that the "individualities" of the molecules in the gaseous state are much better developed than either in the liquid or in the solid state, where they appear to be subject to the collective to a much greater extent. This statement is also in agreement with the short intramolecular distances in the gas phase (p. 30).

Due to the impossibility of registering the individual motions of the molecules in a gas, the *Brown*'ian movement is described as "irregular", "disordered" or "chaotic". It may be noted in this connection that the name "gas" has been derived from the Dutch pronounciation of "chaos" by *van Helmont* (1577 - 1644). He supposed wild motions of the particles, because a gas cannot be kept in an open vessel.

On the other hand, the general gas law expresses ordered relationships on a macroscopic scale. These demand ordered relationships to exist in the submicroscopic areas and hence the *Brown*'ian movement cannot be chaotic. It indicates only our lack of observing and knowing the dynamically ordered relationships.

It has been mentioned before that dynamically ordered relationships require certain conservative boundary conditions. However in the model of the ideal gas all of the static boundary conditions are not considered at all and hence it is not possible to account for the ordered relationships by means of this idealistic model.

This means that the model of the ideal gas concerns a fiction which would neither be observable nor measurable. In order to allow for the mathematical treatment of a real gas on the basis of the ideal gas, the perturbation theory is applied. In this way factors are artificially introduced and physically interpreted as weak interacting forces between the gas molecules and named *van der Waals*-forces. They have a meaning only within the context of the perturbation theory, but they are incorrectly also used for the description of quite different phenomena, such as the adsorption of gas molecules at interfaces. Their characterization by a name led however to the belief that these forces do actually exist, although it is impossible to observe or to measure their contribution to the energy of interaction (p.175).

The other extreme is presented in the model of the ideal crystal, in which perfectly fixed and regularly arranged points are assumed with complete neglect of all dynamic aspects of order. The crystal structure determination ends in an abstraction by which the lattice-geometrical aspects are clearly illustrated by the assessment of the mean values for the lattice parameters. In this way the differentiation of the real crystal has been completely eliminated. Since the differentiation in a crystal is a requirement for all of its properties - and indeed for its observation and for its very existence - this has to be admitted for the description of a real crystal.

Unfortunately, the fiction of the ideal crystal has been assigned the "state of perfect order" and hence all differentiations of a real crystal are considered as deviations from this assumed perfect order (see Chapter 16). Accordingly all aspects of the differentiation are described as "crystal defects", "imperfections" or "crystal faults" and the movements within the crystal as "displacement reactions".

3. Theoretical Approaches

Because a liquid is between these extreme cases, it is impossible to eliminate either the static (conservative) or the dynamic (dissipative) aspects for its description. This means that there is no extreme remaining that could be taken into consideration in an attempt to construct an ideal model for a liquid.

Instead, various approaches have been advanced in order to describe the liquid state, each with its own set of assumptions, approximations and consequently its restricted applicability.

The attempts to describe a liquid as a perturbed solid involves the introduction of holes into the system [1]. The holes are not considered simply as unoccupied regular "cells" but regarded as "fluidized vacancies" that move around in the volume of the liquid or as behaving like gas molecules within the liquid medium.

A significant improvement over the simple hole model was brought about by specifying in a definite way the number and the properties of the holes as well as their degrees of freedom [2,3]. This theory covers certain aspects which are essential for the liquid structure and these will be considered in Chapter 9.

The attempts to characterize a liquid by means of correlation functions are no longer being made because such characterizations may also be applied to the gaseous state [4].

The treatments of a liquid either as a highly densified gas or as a highly perturbed solid have not been successful because of the artificial assumptions and the mathematical complications [5]. It has been stated that our theoretical knowledge about the states of aggregation might be suitable to "invent" both the solid and the gaseous state, but unsuitable to "invent" the liquid state [6]. This has also been expressed as follows [7]: *"Any model for an ideal liquid is bound to be far from reality"*.

All of the attempts to describe theoretically a liquid are characterized by the use of statistical methods. From the statistical point of view, the movements of the molecules are assumed irregular and this assumption is the basis for all kinetic models. The definition of kinetic energy of particles in statistical terms means that *every possible value can be found* and hence all conclusions are in terms of probability functions. These do not allow insight into the actual differentiations and the dynamically ordered relationships.

The statistical method is limited with regard to the achievement of scientific knowledge, since the starting point of the investigation, i. e. the different properties of the parts of the system are not explained. *The statistical method leads to the elimination of the differences, as it ends in the abstraction by assigning equal numbers to different parts.* The individual properties of the parts are lost in this way and they cannot be found in the statistical results. The real differentiation of a system remains disguised in the statistical data as each individual part is accorded the same mean value. Strictly speaking, *statistics begin when reality ends* and therefore they are useful only for the illustration of certain aspects.

4. Suggestions of Ways by which Liquids may be Classified

Because of the enormous variety of liquids of different properties, their classification has been attempted, but all of them are based on certain model assumptions [5].

With regard to the assumed intermolecular forces between the "constituents" the following classification has been suggested:
1. "Regular" liquids (only dispersion forces)
2. Polar and hydrogen bonded liquids (involving dipole and covalent interactions)
3. Electrolyte solutions (described as ion-dipole and ion-ion interactions in a dielectric medium)
4. Molten salts (ionic interactions with no medium)
5. Metallic liquids (electron-ion or electron-molecule interactions)

From the point of view of the differences of the constituent particles the following classification has been favoured:
1. Atomic liquids (liquid inert gases and liquid metals)
2. Molecular liquids (ranging from liquid hydrocarbons to liquid water)
3. Ionic liquids (molten salts)

It has been suggested that molecular liquids may be classified as follows:
1. Non-polar molecular liquids (such as liquid hydrocarbons or liquid chlorine)
2. Aprotic polar molecular liquids (such as acetone or acetonitrile)
3. Hydrogen bonded molecular liquids (such as water or alcohols)

From the point of view of the donor-acceptor approach molecular liquids are classified as follows:
1. Protic solvents (water, alcohols, sulphuric acid, hydrogen fluoride etc.)
2. Aprotic solvents (dimethylsulfoxide, dimethylformamide, acetonitrile, nitromethane, benzene etc.)

It will not be possible in this text to cover all the different kinds of liquids, and thus we decided to consider only molecular liquids and in particular to follow the last mentioned classification.

5. The Molecular Approach to Liquid Water

According to the molecular view, attempts have been made to provide a description of the structure of liquid water by starting from the water molecules [8]. These show, however, different structural characteristics in the gas and in the liquid.

Table 4. Properties of water molecules in the gas and in the liquid

Molecular property	Gas	Liquid
O-H distance	95 pm	99 pm
H-O-H angle	104,5°	109.5°
dipole moment	1.84 D	not measurable
donor number	18	42
acceptor number	50	55

The enormous difficulties in this procedure have been expressed in a textbook as follows [9]: "In spite of the convincing nature of the general qualitative picture and the extensive range of physico-chemical measurements which have been made on water, including in recent years numerous spectroscopic studies, it remains unfortunately true that no satisfactory detailed molecular picture has yet emerged. The present situation is too confused to be briefly summarized, and the same statement is even more true of efforts to give a molecular interpretation of the role of the solvent in solvent-solute interactions in water and other hydrogen-bonded systems".

The association of water molecules in the liquid is described by means of hydrogen bonds (or hydrogen bridges) and an enormous amount of experimental and theoretical studies is available. According to spectroscopic evidence, the hydrogen atoms are not at fixed positions, but rather oscillating between the oxygen atoms of the associated water molecules [10,11]. This means that a clear distinction is not possible between the intramolecular O-H and the intermolecular H-O bond as is made in the theoretical treatments.

According to the donor-acceptor approach the strong association between the water molecules in the liquid is described by their unique ability to act both as a strong donor through its oxygen atom *and* as a strong acceptor through its hydrogen atoms [12] (see p. 231).

This amphoterism allows for the high versatility and the high reactivity of a water molecule and also for the establishment of the three-dimensional network which is extremely adaptable and flexible.

For the formation of a hypothetical dimeric water molecule from the monomeric water molecules the following results have been calculated [13]:

$$+0.0167$$
$$H_{(1)} \searrow \quad -0.0221 \qquad +0.0360 \quad -0.0378$$
$$+0.0167 \nearrow O_{(1)} \longrightarrow H_{(3)} \rightleftharpoons O_{(2)} \searrow -0.0096$$
$$H_{(2)} \qquad\qquad\qquad\qquad H_{(4)}$$
$$+0.0114 \quad -0.0114$$

The formation of the dimeric water molecule may be described by the strong donor function of $O_{(1)}$ and the strong acceptor function of $H_{(3)}$. This leads to a complete reorganization within the developing dimeric unit which is more reactive than its constituent monomeric water molecules. As a result of the $O_{(1)}-H_{(3)}$ intermolecular interaction, the intramolecular $H_{(1)}-O_{(1)}$, $H_{(2)}-O_{(1)}$ and $H_{(3)}-O_{(2)}$ bonds are lengthened, whereas the $O_{(2)}-H_{(4)}$ bond is shortened. In this way the acceptor properties of $H_{(1)}$ and of $H_{(2)}$ are increased and those of $H_{(4)}$ decreased. At the same time the donor properties are improved at $O_{(1)}$ and at $O_{(2)}$, although to a different extent.

The $O_{(1)}-H_{(3)}-O_{(2)}$ sequence is known as hydrogen bond and the $H_{(3)}$ atom is known to oscillate within the distance between the two oxygen atoms continuously, so that a distinction between the bonds $O_{(1)}-H_{(3)}$ and $H_{(3)}-O_{(2)}$ is not possible. Nevertheless, this distinction is made and different bond energies are calculated.

The bond $O_{(1)}-H_{(3)}$ is not considered as a covalent bond since no orbitals seem to be available (modern MO-theory does not suffer from this limitation) and therefore it is regarded as due to *Coulombic* interactions. However, if the electrostatic contribution alone were the major controlling factor in determining the intermolecular hydrogen bond energy, there should be a correlation between its strength and the dipole moment of the electron donating water molecule. This is not the case and it seems therefore, that the covalent contributions are very important in the hydrogen bond.

The actual $O_{(1)}$-$O_{(2)}$ distance depends on the actual environment, which has been artificially eliminated for the consideration of the "isolated" dimeric unit.

The O-H-O bond is symmetrical only, when the O-O distance is very short. In this case the O-H-O bond angle is nearly 180°, the hydrogen bond linear. With increasing O-O distance, the hydrogen bond becomes less symmetrical and an increasing deviation from linearity is observed.

The symmetrical hydrogen bond has a polarizability of about two orders of magnitude larger than usual polarizabilities. The polarizabilities of unsymmetrical hydrogen bonds are somewhat smaller but still considerably greater than those of other bonds [14].

It is hardly possible to determine the optimal configuration for the idealized trimeric unit. Within the monomer $H_{(5)}$-$O_{(3)}$-$H_{(6)}$ reacting with the dimeric unit shown on p. 44, the $O_{(3)}$ atom may act as a donor towards the hydrogen atoms $H_{(1)}$, $H_{(2)}$, $H_{(3)}$ and to a lesser extent $H_{(4)}$, while the hydrogen atoms $H_{(5)}$ or $H_{(6)}$ may attack either $O_{(1)}$ or $O_{(2)}$. The number of possible states in the dimeric unit is further drastically increased because of the oscillations of the hydrogen atoms within the hydrogen bonds, which "causes" all of the other bonds to oscillate. After careful consideration of all possibilities *Richards* [15] came to the following conclusion: "Calculations in a brute force approach have not contributed anything of outstanding significance ... In fact, even a system such as $(H_2O)_3$ has so many possible variables in terms of the positions of the constituent nuclei that it is not possible to do sufficient calculations to satisfy a statistical mechanism without using grotesque amounts of computer time". Faced with these enormous difficulties to predict the structure of an idealized trimeric water unit by starting from three single water molecules, we cannot hope to gain an understanding of liquid water in this way. 3.10^{22} water molecules are contained in one millilitre of liquid water and hence the number of structural possibilities would exceed by far the number of atoms available in the whole universe which has been estimated to be 10^{89}.

6. Structural Models for Liquid Water

Several excellent reviews are available on the structural representations of liquid water [16]. The first attempt to explain the physics of liquid water has been presented by *Bernal* and *Fowler* [17]. They assumed a sort of pseudo-crystalline structure with three different states in properties depending on the temperature, namely

1. water I: tridymite-ice-like, present to a certain degree at low temperatures below 4°C;

2. water II: quartz-like, predominating at ordinary temperatures;

3. water III: close-packed, ideal liquid, predominating at high temperatures for some distance below the critical point at 374°C;

At any temperature the liquid is taken to be homogeneous, its average structure resembling water I, II and III to a greater or lesser degree. The increase in the density of water on melting may thus be ascribed to a transition from the ice I tridymite-like structure to the denser quartz-like structure.

This model has been criticised for its rigidity and its undue bending strain exerted on the hydrogen bonds in the quartz-like structure [18-21], so that it does not account for the fluidity of water, but its principal features, i.e. the existence of extensively hydrogen bonded regions, and the gradual breakdown of hydrogen bonding with rising temperature have been incorporated into most of the later models. They are firmly supported by the results of X-ray scattering experiments and IR, Raman and NMR spectroscopy.

Frank and *Wen* [22] suggested the formation of the hydrogen bonded *net-work* as a cooperative process in which so-called *flickering clusters* of varying extent form, relax and reform in a temporal sequence and in a spatial pattern determined by the energy fluctuations, which are continuously taking place [23]. The changes in local and temporal bond lengths and bond angles are expected in agreement with the later formulated bond length variation rules presented in the preceding chapter. Their continuous changes have recently been illustrated by the results of molecular dynamic simulations.

In the structural models by *Pople* [24] and *Bernal* [25] the hydrogen bonds are considered as variable and flexible, as they are assumed to bend continuously. The free energy of the hydrogen bond is assumed to be a continuous and smooth function of the bond angle. These models are characterized by a long-range order in liquid water, which is highly influenced by the presence of small amounts of solutes.

Several other models have been proposed [16] such as the gas-hydrate model in which the essential role of dissolved gases entrapped in holes and that of unoccupied holes is expressed (Chapter 10).

In general, the models have been divided into two groups [16], namely

1. the *mixture models*, such as the net-work model and the flickering cluster model ,

2. the *continuum models*.

According to *Krindel* and *Eliezer* [16], in the mixture models, water molecules in the liquid may be said to belong at least to two species, i.e. they may be found in at least two different states. The species (states) of water molecules differ in the number of their hydrogen bonds. They are assumed to be short-lived and to interchange roles. The strength, and the bending angles of the hydrogen bonds are considered as discrete quantities with values characteristic of each state. The non-hydrogen bonded species in these models are taken to be much more mobile than the hydrogen bonded species which are situated in rather rigid structures. In order to account for the fluidity of water, these models have to assume a considerable percentage of mobile non-hydrogen bonded molecules.

In the "continuum" models the energy of the hydrogen bond is assumed to be a continuous and a smooth function of the bending angle, so that every bending angle corresponds to a possible temporary equilibrium state of a molecule. There is a continuous distribution of molecular environment in energies. It is also assumed that a high percentage of the hydrogen bonds in water are bent. In a "continuum" model, even the "more-hydrogen bonded" molecules are located in rather flexible structures, so that even if the "less-hydrogen bonded" molecules are postulated to exist in a minute amount, the fluidity of water can be accounted for".

None of the current models of water can be completely ruled out on the basis of thermodynamic and dielectric properties as well as of the results of X-ray scattering measurements of water. Thus, the three-dimensional net-work of water molecules seems to be well-established, but the so called "structural order" in liquid water refers to a *diffusionally averaged structure* with half-life-periods of hydrogen bonded clusters of water molecules in the order of magnitude of 10^{-11} seconds [8].

The term "diffusionally averaged structure" implies that the structural representation has been obtained *in a statistical way by which the actual dynamically ordered relationships have been eliminated.* All efforts by molecular dynamics simulations are based on artificially restricted assumptions and hence they may contribute to illustrate certain aspects without ever reaching the uniqueness of liquid water.

7. Unsolved Problems

The limitations of the molecular concept have been pointed out by *Primas* [27] as follows: "It is indeed impressive how many experts have the absolute faith that, say, the behaviour of liquid water will be reduced to interacting H_2O molecules sometime in the future ... Our vision of the world will be severely limited if we restrict ourselves to the molecular view... If we approach matter from a molecular point of view, we will get molecular answers, and our molecular theories will be confirmed. But different viewpoints are feasible".

It would be an illusion to expect to gain an understanding of nature by consideration from *one* single point of view. From any single point of view shifts and distortions in the projections are unavoidable, so that not even a certain part of the system can be fully described. It is therefore necessary to consider each phenomenon in all respects from different points of view.

In trying to meet this demand, we feel reminded of *Bernal's* statement [28]: "No one, who knows what the difficulties are, now believes that the crisis of physics is likely to be resolved by any simple trick or modification of existing theories. Something radical is needed, and will have to go far wider than physics. A new world outlook is being forged, but much experience and argument will be needed before it can take a definite form. It must be coherent, it must include and illuminate the new knowledge and it must have a different dimension from all previous world views".

We believe that scientific knowledge is not complete, unless the phenomena are fully understood. We should no longer deny or eliminate from our investigation those phenomena, for which our models provide no explanation. We should no longer take for granted explanations, which have been given on the basis of one of the model assumptions, because these allow the investigation only of certain minimum conditions.

For example, a real crystal is described according to the minimum conditions of its idealised crystal structure. In this way *the actual differentiation, which is characteristic for the real crystal, is statistically eliminated.* Although in this way certain pictures of the crystal structure are well illustrated, it is impossible to apply an analogous approach to liquid water because of the absence of a model for an ideal liquid.

Because water has a central position for all phenomena in nature, we ought to use only such concepts which are in agreement with its "true nature". We must allow our intellect to *"read into the things"*, to be guided by the things as they are and to reach conceptions which are adequate to them. We should no longer allow our intellect to be guided by the artificial frameworks of model assumptions or theories.

This requires an approach which is in agreement with the demand expressed in ancient philosophy as *"Adaequatio rei et intellectus"*.

These considerations show clearly that a broad philosophical background is necessary for scientific knowledge. It is, however, a fact that many scientists are suspicious of philosophy and it is certainly true that a scientist can hardly be expected to be a good philosopher and scientist at the same time.

To this problem *Goethe* [29] made the following comment: "One cannot expect the physicist to be a philosopher, but one can expect him to possess enough knowledge of philosophy to be able to distinguish between himself and the world and to unite himself with it again on a higher level".

The present attitude of many scientists towards philosophy has been expressed by *Weizsäcker* [30] as follows: "It is an empirically proved fact that almost all the theoretical physicists of our time practise philosophy ... In contact with the object of his research, the physicist develops thought patterns which are adapted to his subject, but do not always conform to the traditional methods of philosophy. He senses this and acquires an instinctive aversion to philosophy, since he knows it only as a commodity sold to him by a suspicious merchant which he feels he cannot use in his household in its present form. At the same time, when he makes use of terms such as truth, reality, nature, phenomenon etc., to explain to himself and to others what he is doing and what he wants to express, he does not realize that every word he uses is already a piece of philosophy and that with his rejection of the philosopher he hasn't got rid off philosophy, but has himself become a philosophic dilettante. Unconscious philosophy is, however, on the whole of a worse quality than conscious philosophy, and so it is precisely the most profound thinker of modern physics who returns to philosophy on his own".

This philosophy is, however, usually concerned with the results of scientific investigations rather than with nature as it is [31].

We have therefore to bear in mind that science is the result of a certain attitude of man towards nature, the result of a certain attitude of the investigator. The result depends on the terms used, the questions asked and the methods employed. Quantum physics has demonstrated the impossibility of the separation of investigator and object to be investigated.

This problem has been commented by *Schrödinger* [32] as follows: "Without meaning to, sometimes without realizing that he is doing so, the scientist tries to simplify his problem of understanding nature by leaving his own person, the "perceiving" subject, out of the desired view of nature. Almost without noticing it, the thinker steps back into the role of the passive observer. This is an enormous facilitation of his task ... This important step of leaving oneself out of stepping back into the position of an onlooker who has nothing to do with the performance, has been given names that make it seem quite harmless, natural and inevitable. One could simply call it objectivizing, regarding the world as an object. The moment this is done, one has effectively excluded oneself. And the reason for this intellectual embarrassment is just that, in order to formulate a personal view of the world outside ourselves, we have allowed ourselves the extraordinary simplification of removing our own person to, so to speak, "cut it out" of the picture. Especially - and this is the most important point - this is the reason why the scientific view of the world does not in itself contain any ethical and aesthetical values, nothing about our own destination, our ultimate purpose and - pardon me - no God."

In contrast to the sciences that study a certain clearly defined and limited field, philosophy searches for universality and totality, for universally valid statements and answers. This explains why, no matter how and how far this requirement is or can be fulfilled, philosophy must accompany every scientific investigation from the beginning to the end.

It has been emphasized that it is the experience of reality that paves the way for thought and imagination, but modern science has placed the results of thought before and above experience. In this way reality is no longer investigated in all its dimensions, but only according to the simplified conceptions that we have made about the things (for ourselves). Such ideas seem to be perfect and transparent because they do not require matter and consequently have no uncertainties.

One is tempted to value the idea more highly than reality because the idea is the result of a transference of natural phenomena from the material into the intellectual sphere. Although it cannot reach reality, it can be used to illustrate certain aspects of nature. The idea is, therefore, imperfect with regard to nature, but nature seems imperfect as long as the idea is considered as perfect. This misunderstanding may be demonstrated by the scientific description of a real crystal as "imperfect" with regard to the idea of the "perfect" model of the ideal crystal. With this attitude the researcher allows himself be guided by the perfection of ideas and places himself above nature itself.

In this way he encounters the danger of seeing and judging reality through the rose-tinted glasses of ideas and of taking notice only of what "fits in" with them, thereby missing the great variety of reality's manifestations. At the same time he wins, however, a certain independence of reality, which seems necessary to the attainment of certain wished-for, isolated effects: he puts a higher value on his own intentionality than on reality.

The danger of mistaking the ideas for the reality has been realized by *Boltzmann* [33], when he wrote: "I called theory a purely spiritual inner image, and we have seen what a high degree of perfection it can attain. In perfecting the theory even more, how can he avoid to take the image for the reality itself? ... Thus, it can happen to the mathematician that - constantly occupied with the formulas and dazzled by the perfection in themselves - he begins to regard their interrelations as that which really exists and turns away from the real values. The poet's complaint will then be this: that his works have been written with his heart's blood and that supreme wisdom verges on supreme madness".

The usefulness and the limitations of models have been expressed by *Kac* [34] as follows: "Models are for the most part caricatures of reality. But if they are good, then, like good caricatures, they portray, though perhaps in distorted manner, some of the features of the real world".

If we try to take as the point of departure reality in its apparent complexity rather than the idealised models (see p. 171), we are, however confronted with the fact that from this point of view acquisition of scientific knowledge is difficult and vulnerable. However, it appears unavoidable to choose this starting point if an understanding of nature is sought (see Chapter 16).

We have to start with the collection of all phenomena obtained by observations and measurements which *Aristotle* calls the *elements of cognition*.[1]

The next step is the search for the causes and principles by means of the process of induction, which, however, is the task of the philosopher rather than that of the scientist. He applies the causes and the principles, found by the philosopher, for ordering all findings with regard to the natural relationships.

When we start in the following chapters with the first step and try to collect the wealth of phenomena, known about liquid water, especially those under natural conditions, this procedure may be considered as unscientific, or even boring. It may also be questioned because it is believed impossible to encompass the enormous amount of observations and measurements on liquid water and its solutions, but we shall see the further procedure in Chapter 16.

In the chapters to follow we shall try to provide a short outline of phenomena in complex situations. We suggest to collect facts and trivia about water and its solutions, followed by a short account about other molecular liquids, before we can proceed in order to gain an appropriate *understanding of the phenomena*, which will be outlined in Chapters 16 to 21.

References

1. H. Eyring, *J. chem. Phys.* **4** (1936) 283.
2. H. Eyring and M. K. John, *Significant Liquid Structures*, (Wiley, New York, 1969).
3. Y. Marcus, *Introduction to Liquid State Chemistry*, (Wiley, London, New York, 1977).
4. U. Müller-Herold, *Chimia* **39** (1985) 3.

---

[1] According to *Aristotle*, science is the complete cognition of elements, causes and principles. Every act of cognition starts from the elements. Every object, every phenomenon, every event is regarded as an element which must be judged and investigated with the help of causes and principles. A principle is the fundamental, intellectual realization of things. A principle has been attained, if with its help, everything can be understood and judged. However, principles are not suitable for the complete and precise description of details, but suitable for orientation in the arrangement of experiences within their natural relationships, as well as for asking well-aimed questions (see Chapter 16).

5. J. E. Lennard -Jones and A. F. Devonshire, *Proc. Roy. Soc. London (A)*, **163** (1937) 53, **169** (1938) 3/7 and **170** (1939) 464.

6. K. F. Herzfeld and M. Goeppert-Mayer, *J. chem. Phys.* **2** (1934) 38.

7. G. Jäger, *Ann. Phys.*, **11** (1903) 1077.

8. E. Eisenberg and W. Kauzmann, *The Structure and Properties of Water*, (Clarendon Press, Oxford, 1969).

9. J. C. Bailar, H. J. Emeléus, R. S. Nyholm and A. F. Trottmann-Dickenson, *Comprehensive Inorganic Chemistry*, Vol. 1, p. 137, (Pergamon Press, Oxford, 1973).

10. P. Schuster, G. Zundel and S. Sandorty, (eds.), *The Hydrogen Bond - Recent Developments in Theory and Experiment*, (North Holland Publ. Co., Amsterdam 1976).

11. F. Franks in: *Hydrogen Bonded Systems*, eds. A. K. Covington and P. Jones, (Taylor and Francis, London 1968).

12. V. Gutmann, *The Donor-Acceptor Approach to Molecular Interactions*, (Plenum Press, New York 1977).

13. D. Hankins, J. M. Moskowitz and F. E. Stillinger, *J. chem. Phys.* **53** (1970) 4544.

14. R. Janoschek, E. G. Weidemann, H. Pfeiffer and G. Zundel, *J. Am. Chem. Soc.* **94** (1972) 7384.

15. W. G. Richards, in: *Water a Comprehensive Treatise*, Ed. F. Franks, Vol. 6, p. 123 (Plenum Press, New York, 1977).

16. P. Krindel and I. Elizier, *Coord. Chem. Rev.* **6** (1971) 217.

17. J. D. Bernal and R. H. Fowler, *J. chem. Phys.* **1** (1933) 515,

18. A. Eucken, *Z. Elektrochem.* **52** (1948) 255.

19. J. A. Pople, *Proc. Roy. Soc. (London), Ser. A*, **205** (1951) 163.

20. F. S. Feates and D. J. G. Ives, *J. Chem. Soc.*, (1956) 2798.

21. L. Pauling, in *Hydrogen Bonding*, ed. D. Hadzi, p. 1 (Pergamon Press London, 1959).

22. H. S. Frank and W. Y. Wen, *Disc. Farad. Soc.* **24** (1957) 133.

23. H. S. Frank and A. S. Quist, *J. chem. Phys.* **34** (1961) 605.

24. J. A. Pople, *Proc. Roy. Soc. London, Ser. A*, **205** (1951) 163.

25. J. D. Bernal, *Proc. Roy. Soc. London, Ser. A*, **280** (1964) 299.

26. H. M. Chadwell, *Chem. Rev.* **4** (1927) 375,

27. H. Primas, *Chimia* **36** (1982) 293.

28. J. D. Bernal, *Wissenschaft in der Geschichte*, p. 963, (VEB-Verlag, Berlin, 1961).

29. J. W. Goethe, *Zur Farbenlehre*, I. Band, Polemischer Teil (Cotta'sche Buch-handlung Tübingen, 1810).

30. C. F. v. Weizsäcker, *Die Tragweite der Wissenschaft*, p. 201 (Hirzel, Stuttgart 1984).

31. R. Löw, *Philosophia Naturalis* **22** (1985) 343.

32. E. Schrödinger, *Nature and the Greeks*, Shearman Lectures, University College, London, 1948).

33. E. Broda, *Ludwig Boltzmann, Mensch, Physiker, Philosoph,* p.108 (VEB-Verlag Berlin 1957).

34. M. Kac, *Science* **166** (1969) 665.

CHAPTER 6

ANOMALOUS PHYSICAL PROPERTIES OF LIQUID WATER

1. General

Water is unique in almost all of its properties. It has not been possible so far to include water in any general scheme or to account for the great number of its striking properties by any simple theory.

Liquid water is probably one of the least understood liquids, although some of its physical properties have been accepted as international standards, e.g., triple point, boiling point, melting point, density, viscosity.

Many of its properties are called "anomalies" and these have been extensively described by *Dorsey* [1] in his book entitled "Properties of Ordinary Water Substance". This title implies that the anomalies are characteristic properties of ordinary water and seem to be necessary for the irreplaceable role of water in its hydrological cycle and in living organisms.

The term anomalies indicates shortcomings of our interpretations rather than inadequacies of nature [2]. It is therefore necessary to pay special attention to them precisely because they do not meet expectations from theoretical considerations and because they may contribute towards a gain in the understanding of the unique properties of liquid water. One should not even entirely ignore those results which are not strictly reproducible, but rather to try to find the natural reasons for poor reproducibilities by including all available observations and to order them carefully with regard to the natural requirements.

2. Anomalies in the Normal Liquid Range

Melting Point and Boiling Point

The unusually high values for the melting point and for the boiling point have been explained by unusual molecular properties, namely the strong association between water molecules by means of hydrogen bonds.

Although hydrogen bonding is stronger in liquid hydrogen fluoride than in liquid water, its melting and boiling point is considerably lower (Table 5). Its properties as a solvent are less developed, but similarities in solubility properties to water are apparent. For example, carbohydrates and proteins, including the fibrous proteins, which are hardly soluble in water, are readily dissolved in hydrogen fluoride and may be recovered from its solutions [3]. Examples are ribonuclease, tyrosine, serum globulin, collagen, haemoglobin, insulin, chlorophyll and vitamin B12, the latter giving a deep olive-green solution in hydrogen fluoride.

The crucial difference between liquid water and liquid hydrogen fluoride is the different structural arrangement. Because the hydrogen fluoride molecule can form only two hydrogen bonds, namely one by the acceptor properties of the hydrogen

atom and one by the donor properties of the fluoride ion, zig-zag chains and rings
are present in liquid hydrogen fluoride.

Table 5. Liquid range and hydrogen bond energies of several liquids

Species	m.p.[°C]	b.p.[°C]	Critical point [°C]	Crit.pressure [atm.]	H - bond energy [kJ.mol^{-1}]
CH_4	- 182.5	- 161.5	- 82.5	45.8	-
NH_3	- 77.8	- 33.4	+ 132.5	112.5	12
OH_2	0	+ 100	+ 374	218.3	20
FH	- 83.7	+ 19.5	+ 230	64	29
CH_3OH	- 97	+ 64.7	+ 240	78.5	19
C_2H_5OH	-114	+ 78.4	+ 243	63	20
n - C_3H_7OH	-126	+ 97.2	+ 264	51	11

*In contrast to the one-dimensional association in liquid hydrogen fluoride, a
three-dimensional network is established in liquid water,* because each water
molecule can form four hydrogen bonds: two by the acceptor properties of each of
its hydrogen atoms and two by the donor functions of the oxygen atom. This
network is extremely strong and yet extremely flexible and adaptable (Chapter 5).

Liquid ammonia is associated by weaker hydrogen bonds than liquid water
(Table 5) and one-dimensional aggregates are established. Liquid ammonia shows
excellent properties as a solvent, although it is in many ways less reactive than
water. Whereas alkali metals are vigorously oxidized by water to give alkali
hydroxides and hydrogen, deep blue solutions are formed in liquid ammonia, which
contain solvated electrons and hence these solutions show unusually high reducing
properties.

Methanol and ethanol are associated by hydrogen bonds with hydrogen bond
energies similar to those in liquid water. The alcohol molecules differ, however, from
water molecules by forming only three hydrogen bonds. This means that by addition
of one of these alcohols to liquid water, its structural network is somewhat loosened,
but otherwise substantially maintained and these features will be treated in more
detail in Chapter 12.

In the absence of the anomalously high melting- and boiling point, the liquid
range of water at atmospheric pressure would be somewhere between -80°C and
-110°C. Under these circumstances no liquid water would be on earth and neither
life nor the hydrological cycle would be possible.

Heat of Vaporization and Surface Tension

The strong bonding between the water molecules in the bulk liquid requires
strongly bonded molecules at the interface. This is reflected in anomalously high
values for the heat of vaporization and of surface tension. The latter is two or three
times higher than for other liquids. It may be mentioned that it is difficult to obtain

precise values for the surface tension because "impurities" in the range of parts per billion affect the experimental results [5]. As a result of the high surface tension, water penetrates into rock crevices, where, on freezing, it fragments the rock substances.

Density

The density maximum of liquid water at +4°C is well-known and has been explained by the presence of different "pseudocrystalline" forms, varying in extent with temperature [6,7]. Owing to this anomaly, water exposed to low temperature freezes only at the surface, liquid water sinks as it reaches +4°C and forms a layer beneath the upper crust of ice, through which heat is only slowly transmitted.

In the absence of this anomaly, ice would sink to the bottoms of polar seas, lakes, and rivers. Instead of water being insulated from further freezing by a surface layer of ice, the ice insulated by a surface layer of water would be prevented from melting. It would gradually increase in volume, until all aqueous ecosystems that are ice-covered in winter became ice and life in them would not be possible.

Less-known are the consequences of this density maximum for the circulation within seawater with biological importance for the deep waters. Water cooled in the North Atlantic to the temperature of maximum density sinks - fully charged with oxygen - to the ocean floor and spreads out to all the oceans of the world.

Heat Capacity

The heat capacity of liquid water is about twice that of ice and this means that water gains or loses a large amount of heat before its temperature is appreciably changed.

The heat capacity of water shows an anomalous temperature dependence: at 0°C and at 100°C it is slightly greater than at +15°C and shows a flat minimum at +37.5°C [7] (Fig 4).

Minimum values in specific heat and in entropy change have been found to occur in the course of a gradual phase transition in liquid alloys [8]. This is in contrast to the findings for phase transitions in the solid state, where maximum values are observed for both the heat capacity and the entropy changes.

The flat minimum in heat capacity in liquid water at +37.5°C may be an indication for a gradual phase transition in liquid water at this temperature. Such considerations have been advanced by *Trincher* [7], who considers three components in liquid water in the temperature range between 0°C and +60°C, namely the R-component (ice relics), the K-component (the quasi-crystalline water) and the F-component (the liquid water). According to Fig. 5 only the R-component and the K-component are present in the temperature range between 0°C and +15°C. Between +15°C and +30°C the F-component is dispersed in the K-component, while between +45°C and +60°C the K-component is dispersed in the F-component. In the temperature range from +30°C to +45°C gradual transitions between them take place.

The temperature of +37.5°C indicates the optimal penetration of the K- and the F- phases and hence the gradual phase transition between them. This means that at

+37.5°C - the body temperature of the mammals - liquid water shows the greatest differentiation.

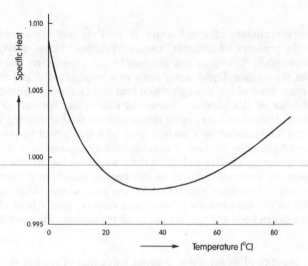

Fig. 4. Heat capacity of liquid water as a function of the temperature.

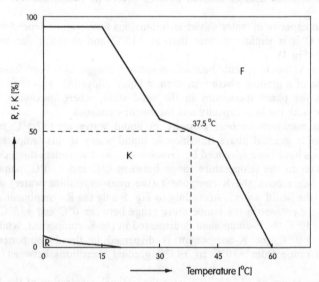

Fig. 5. Phases in liquid water according to *Trincher* [7].

The large heat capacity of water allows the body to maintain nearly isothermal conditions and the oceans to act as large thermostats.

Thermal Conductivity

Whereas for most liquids the thermal conductivity falls with increasing temperature, for liquid water it rises approximately linearly between +20°C and +60°C and passes through a maximum at about +130°C, which has not been explained [9].

Compressibility

Liquid water exerts an exceptionally high resistance towards pressure. At ordinary temperature the coefficient of compressibility is only 5.10^{-1} bar^{-1}. Even more anomalous is the temperature dependence of the coefficient of compressibility, which is characterized by a minimum value at +45°C [10, 11], (Fig. 6).

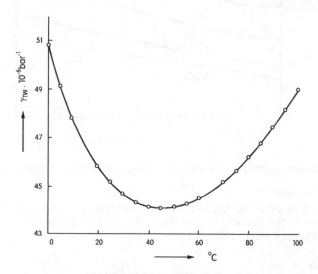

Fig. 6. Temperature dependence of the compressibility of liquid water.

Viscosity

The high viscosity of liquid water has been related to the toughness of the interlocked hydrogen bonded structure. An interesting anomaly is provided by the pressure dependence of the viscosity of liquid water. In all other liquids the viscosity is increased as the external pressure is increased, but in liquid water the viscosity is decreased; below 35°C water becomes more fluid, as the pressure is moderately increased [12]. At temperatures below + 35°C, the pressure-viscosity curves passe through a minimum (Fig 7).

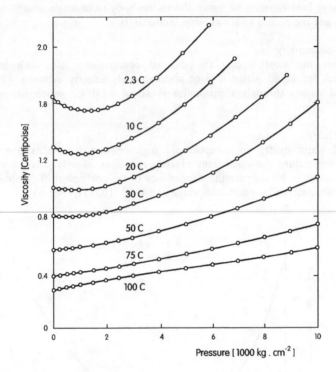

Fig. 7. Pressure dependence of the viscosity of liquid water at different temperatures [12].

Dielectric Constant

Compared to other liquids, the dielectric constant of water is high. For this reason its role in the attempts to provide an electrostatic interpretation of solution phenomena has been highly overrated.

Other Effects

Drost-Hansen [13] reported "kinks" and "thermal discontinuities" and suggested transitions to occur in liquid water at +15°C, +30°C, +45°C and +60°C.

3. Anomalies in the Supercooled State

Temperature Range of Supercooled Water

The conditions under which supercooled water can be obtained have been extensively investigated. Experiments were made (i) in glass, quartz and silica capillaries of various diameters (range between 10 μm up to test-tube size [14 -23]) partly coated by hydrophobic substances [20], (ii) on droplets placed on different solid surfaces (ranging from 3 μm to 50 μm) [24-28], (iii) on suspensions of droplets at the interface of inmiscible liquids (mean diameter between 50 μm and 2,5 cm) [27-34], (iv) on emulsions (diameter of droplets of about 4 μm) and (v) in the cloud chamber (droplet diameters below 1 μm) [31-34].

The freezing points obtained as a function of the mean diameter of the capillary tubes are shown in Fig. 8 and those as a function of the mean diameter of the droplets in Fig. 9. The data are difficult to compare, because different experimental conditions were used (nature of the adjacent phase, purity of the sample and cooling mode).

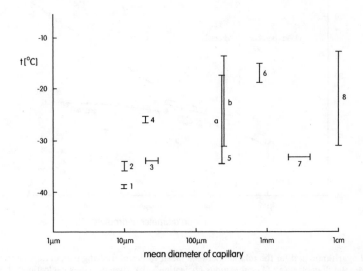

Fig. 8. Logarithmic plot for freezing temperatures of supercooled water in different capillaries in relation to their diameters; 1: glass capillaries (10 μm), 2: capillaries (10 μm), 3: Pyrex capillaries (25 μm), 4: Pyrex capillaries (20 μm), 5a: glass tubes coated by hydrophobic substances (250 μm), 5b: uncoated glass tubes (250 μm), 6: glass tubes (0,8 mm), 7: quartz tubes (2-4 mm), 8: glass tubes (1 cm).

With regard to the purity of the sample, kinetic interpretations have been provided. A distinction has been made between "heterogeneous nucleation" and "homogeneous nucleation" (Fig. 9) [20,29].

Within each of these groups the liquid range that can be obtained is greater the smaller the diameter of the capillary tube and the smaller the drop size. The lowest temperature which has ever been obtained in capillary tubes is -39°C [21], in emulsions -40°C and in cloud chamber experiments -44°C [31,32]. It has further been shown that in a capillary tube of given diameter the freezing point is lowered as the capillary is coated by a hydrophobic layer.

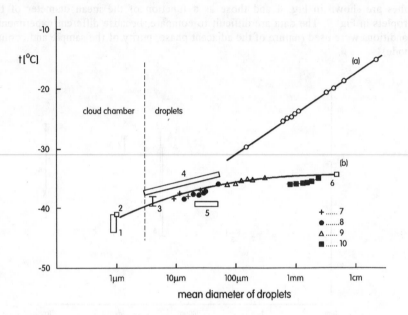

Fig. 9. Logarithmic plot for the relation between mean values of freezing temperatures and mean droplet diameter; (a): "heterogenious nucleation"; (b): "homogenious necleation"; 1: cloud chamber experiments (< 1 μm); 2: cloud chamber experiments (about 1 μm); 3: freezing of an emulsion of water in *n*-heptane + sorbitan tristearate (mean particle size 4 μm); 4: droplets condensed on colloidal films supported by highly polished metal surfaces (3 - 50 μm); 5: emulsion in silicon oil (20 - 50 μm); 6: water drops (4 - 6 mm) suspended between mercury and silicon liquid; 7: droplets condensed on a metal surface and covered with silicon oil; 8: water droplets condensed on a silicon oil surface and covered with liquid paraffin; 9: droplets at the interface of liquid paraffin and carbon tetrachloride; 10: analogous to 9.

Molar Heat Capacity

It has been mentioned that the heat capacity of liquid water at atmospheric pressure increases slightly below +37.5°C as the temperature is lowered [35], but below 0°C it increases dramatically as the temperature is further decreased [36]. Fig. 10 shows that at -35°C the heat capacity is by 34.7% higher than that of liquid water at its boiling point.

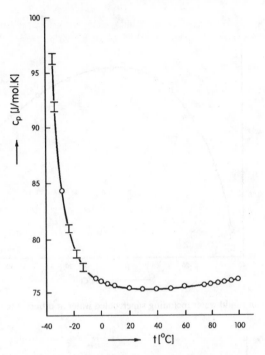

Fig. 10. Heat capacity of supercooled water from emulsion measurements (water in *n*-octane) [36].

Wilhelm et al. [37] calculated the difference between the partial molar heat capacity of oxygen and nitrogen respectively in aqueous solutions at infinite dilution, and for the ideal gas state for different temperatures. The results showed that the heat capacities of these gases are higher when dissolved in water than in the gaseous state.

Density

Measurements of the density of supercooled water have been made in the temperature range from 0°C to -34°C [39]. Fig. 11 shows that the density is drastically decreased as the temperature is lowered. At -34°C the density of supercooled water is almost the same as at +70°C (Fig. 11).

Compressibility

Results of the measurements down to a temperature of -26°C [23] show that the increase in compressibility is greater the lower the temperature (Fig. 12). At -26°C the isothermal compressibility for liquid water is by 45 % higher than at 100°C!

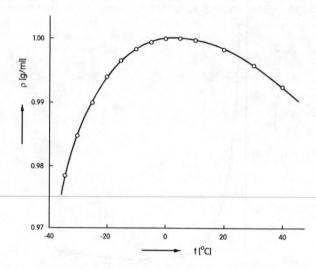

Fig. 11. Density of liquid water including supercooled water at different temperatures [39].

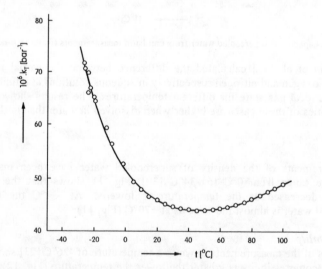

Fig. 12. Temperature dependence of the isothermal compressibility k_T of liquid water between -35°C and +100°C [23,35].

Gas Solubilities

It is well-known that within the normal liquid range of water gas solubilities show a negative temperature coefficient which increases with decrease in temperature. Measurements on the oxygen content of supercooled water have shown that this tendency is continued, i.e. the oxygen content rises increasingly as the temperature is lowered [38] (Fig. 13).

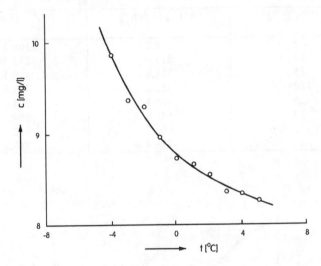

Fig. 13. Oxygen content in supercooled water [38].

Free Energy and Entropy

For a given temperature, values of the free energy difference between water and ice, ΔG^0, have been calculated [40] by extrapolation of the difference between the partial molar free energy of water in aqueous solution and the molar free energy of ice at the same temperature to zero concentration. The free energy difference between supercooled water and ice for water supercooled in capillaries ΔG^0_{cap} and in emulsions ΔG^0_{em} were calculated from c_p-data [36, 41-43] in the temperature range from 0°C to -40°C (Table 6). The results show that

(i) at a given temperature the free energy of supercooled water is higher than that of ice,

(ii) the free energy differences between water supercooled in capillaries and in emulsions become greater the lower the temperature [41].

The comparison of ΔG^0_{cap} and ΔG^0_{em} reveals that the free energy difference of water in emulsion is consistently higher than that in "bulk" water [40]. The differences in free energy between water in emulsions and in capillaries are greatest between -20°C and -30°C (Fig. 14).

The partial derivative of the free energy differences ΔG^0_{em} and ΔG^0_{cap} respectively according to temperature provides the values for the entropy differences ΔS^0_{em} and ΔS^0_{cap}. The plots of these values vs. temperature (Fig. 14) show that the entropy differences decrease when the temperature is lowered.

Table 6. Free energy differences between supercooled water and ice at different temperatures for water supercooled as emulsion and in capillaries respectively [40].

t[°C]	ΔG^0_{em} [J.mol^{-1}]	ΔG^0_{cap} [J.mol^{-1}]
- 5	108.2	108.1
-10	213.0	212.5
-15	314.4	313.1
-20	412.3	409.6
-25	506.3	501.7
-30	595.8	589.3
-35	680.2	672.1
-40	758.4	749.8

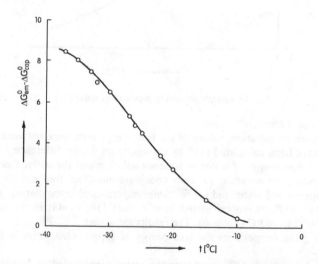

Fig. 14. Differences between free energy of water in emulsions and capillaries in the supercooled region [40].

In the temperature range between - 20°C and -30°C the entropy difference between water in emulsions and in capillaries is greatest (Fig.15). At given temperature in the supercooling region the values of ΔS^0_{em} are always greater than the values of ΔS^0_{cap} [41].

Angell concluded his review on supercooled water as follows [42]: "It is evidently appropriate to conclude that while the largely recent study of water in the metastable, supercooled state has introduced new and suggestive relationships, it has also produced buffling new questions and new demands which must be satisfied by theoretical models of this most important and perplexing of all liquids".

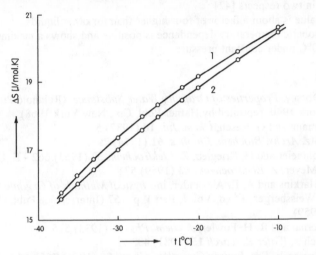

Fig. 15. Differences between entropy of supercooled water and ice vs. temperature [40]. Curve 1 refers to water droplets in emulsions, curve 2 to water in capillaries.

4. Liquid Water at High Temperatures and Pressures

Knowledge on the thermodynamic properties of water at high temperatures and pressures have become important in designing power stations, in corrosion problems, in growing synthetic crystals from hydrothermal solutions and in understanding many geochemical processes [43].

The liquid range of water can be extended above +100°C at atmospheric pressure, when - apart from the temperature - similar conditions are provided as they are needed to obtain supercooled water [44]. Highly purified water has been superheated in capillaries up to +240°C [45] or in form of small drops in a hydrophobic liquid such as benzyl benzoate up to +239.5°C [46].

When this limit of temperature is reached, the liquid remains only for a few seconds before an explosion takes place. In the presence of 0.6% sulphuric acid the maximum superheating temperature in capillaries is only +213°C, with subsequent explosion within two or three seconds [47].

The liquid range is extended to higher temperatures as the external pressure is increased. The boundary conditions for the liquid state are characterized by the critical temperature (for water +374°C) and the critical pressure (for water 218atm).

Under these conditions the surface energy vanishes and hence a phase boundary can no longer be maintained.

In the temperature range from $+100°C$ to $+1000°C$ and the pressure range of 1 to 250 kbar the specific volume of water is a monotonous function of temperature and pressure. It does not show anomalies, whereas the thermal conductivity is anomalous in two respects [42]:

1. Its value is about a factor of four higher than for other liquids,
2. its isobaric temperature dependence is positive and shows a maximum around $+200°C$ under constant pressure.

References

1. N. E. Dorsey, *Properties of Ordinary Water Substance*, (Reinhold Publ. Corp. New York 1940, reprinted by Hafner Publ. Co., New York 1968).
2. V. Gutmann and G. Resch, *Chem. Int.* **10** (1987) 5.
3. J. J. Katz, *Archs. Biochem. Biophys.* **61** (1954) 293.
4. W. Strohmeier and G. Briegleb, *Z. Elektrochem.* **57** (1953) 662 - E. U. Franck and F. Meyer, *Z. Elektrochem.*, **63** (1959) 571.
5. W. D. Harkins and A. E. Alexander, in *Physical Methods of Organic Chemistry*, Ed. A. Weissberger, 3rd ed. Vol. I, Part I, p. 757 (Interscience Publ. Co., New York 1959).
6. J. D. Bernal and R. H. Fowler, *J. chem. Phys.* **1** (1933) 515.
7. K. Trincher, *Water Research* **15** (1981) 433.
8. K. L. Komarek, *Ber. Bunsen Ges. phys. Chem.* **80** (1976) 765 - **81** (1977) 936 - A. Janitsch, K. L. Komarek and J. Mikler, *Z. Metallk.* **71** (1980) 629.
9. R. W. Powell, *Adv. Phys.* **7** (1958) 276.
10. G. G. Kell, *Chem. Engng. Data* **12** (1967) 66.
11. R. A. Fine and F. J. Millers, *J. chem. Phys.* **59** (1973) 3529.
12. K. E. Bett and J. B. Cappi, *Nature* **207** (1965) 620.
13. W. Drost-Hansen, *Ind. Eng. Chem.* **61** (1969) 10.
14. J. A. Schufle and M. Venugopalan, *J. Geophys. Res.* **72** (1967) 3271.
15. N. Muller and J. A. Schufle, *J. Geophys. Res.* **73** (1968) 3345.
16. J. A. Schufle and Nai Teng Yu, *J. Colloid. Interface Sci.* **46** (1968) 395.
17. B. V. Zheleznyi, *Russ. J. Phys. Chem.* **43** (1969) 1311.
18. G. Tammann and A. Büchner, *Z. anorg. allg. Chem.* **222** (1935) 371.
19. R. G. Wylie, *Proc. Phys. Soc.* **B 66** (1953) 241.
20. S. C. Mossop, *Proc. Phys. Soc.* **B 68** (1955) 193.
21. R. S. Chahall and R. D. Miller, *Brit. J. Appl. Phys.* **16** (1965) 231.
22. J. Meyer and W. Pfaff, *Z. anorg. allg. Chem.* **224** (1935) 305.
23. R. J. Speedy and C. A. Angell, *J. chem. Phys.* **65** (1976) 851.
24. G. H. Pound, L. H. Madonna and S. Peake, *J. Colloid. Sci.* **8** (1953) 187.
25. W. Jacoba, *J. Met.* **12** (1955) 408.
26. A. E. Carte, *Proc. Phys. Soc.* **B 69** (1956) 1028.
27. E. K. Bigg, *Proc. Phys. Soc.* **B 66** (1953) 688.
28. E. K. Bigg, *Quart. L. Met. Soc.* **79** (1953) 510.

29. E. J. Langham and B. J. Mason, *Proc. Roy. Soc. London* **A 247** (1958) 493.
30. H. Bayardelle, *C. R. Acad. Sci,* **239** (1954) 988.
31. D. M. Anderson, *J. Colloid. Interface Sci.* **25** (1967) 174.
32. D. M. Anderson, **216** (1967) 563.
33. B. M Cwilong, *Proc. Roy. Soc. London* **A 190** (1947) 137.
34. B. J. Mason, *Clouds, Rain and Rainmaking,* (Cambridge Univ. Press, 1962).
35. *Handbook of Chemistry and Physics,* 67th ed. (CRC-Press, Boca Raton, 1986/87).
36. C. A. Angell, M. Oguni and W. J. Sichina, *J. phys. Chem.* **86** (1982) 998.
37. E. Wilhelm, R. Battino and R. J. Wilcock, *Chem. Rev.* **77** (1977) 219.
38. G. Scheiber and V. Gutmann, *Monatsh. Chem.* **124** (1993) 277.
39. D. E. Hare and C. M. Sorensen, *J. chem. Phys.* **84** (1986) 5085.
40. J. V. Leyendekkers and R. J. Hunter, *J. chem. Phys.* **82** (1985) 1440.
41. V. Gutmann, E. Scheiber and G. Resch, *Monatsh. Chem.* **120** (1989) 671.
42. C. A. Angell, in *Water, a Comprehensive Treatise,* ed. F. Franks, Vol. 7, p.1, (Plenum Press, New York, 1979).
43. K. Tidlheide, in *Water, a Comprehensive Treatise,* ed. F. Franks, Vol. 1, chapter 13, (Plenum Press, New York, 1972).
44. V. Gutmann, G. Resch und E. Scheiber, *Rev. Inorg. Chem.* **11** (1991) 295.
45. M. v. Stackelberg and H. R. Müller, *Z. Elektrochem.* **58** (1954) 25.
46. R. E. Apfel, *Nature, Phys. Sc.* **238** (1972) 63.
47. F. R. Kenrick, C. S. Gilbert and K. L. Wismer, *J. phys. Chem.* **28** (1924) 1297.

CHAPTER 7

SOME TRIVIA ABOUT WATER

1. General

Studies of aqueous solutions have contributed enormously to the development of chemistry: inorganic chemistry, analytical chemistry, physical chemistry, electrochemistry and coordination chemistry have been advanced in this way. Quite different aspects of water have, however, been disclosed by other branches of science, such as biochemistry, biology, ecology, pharmacology, medicine, oceanology, geochemistry, geology or modern technologies.

If a full understanding of water is intended, all of these results ought to be considered and appreciated. Unfortunately, it is obviously impossible to provide an appropriate account about the enormous versatility of liquid water. An expert in a certain field cannot be expected to have sufficient background knowledge in other fields.

For this reason the authors of the book cannot hope to provide an account of the various aspects of water in nature and the reader of the book can hardly be expected to appreciate the enumeration of many well-established facts about water.

We ought, however, to bear in mind that all materials on earth seem to respond to the actions of water in one way or another. Many materials are readily dissolved and many others to a slight but detectable extent. Others are decomposed by water or attacked but slightly and slowly. Liquid water is one of the most corrosive liquids known and yet it is, unlike other liquids, physiologically innocuous.

2. Water as a Solvent

Solubilities range from complete miscibility with water in all proportions to nearly insoluble compounds, such as HgS. Solutes may be classified as

1. Hydrophilic ("water-attracting") solutes
2. Hydrophobic ("water-repellent") solutes
3. Amphipathic (aggregating) solutes.

Complete Miscibilities

Water is miscible in all proportions, in the first place, with hydrogen bonded liquids, such as hydrogen fluoride, nitric acid, sulphuric acid, methanol, ethanol, *n*-propanol, *t*-butanol, glycol (1,2 ethane-diol) or glycerol (1,2,3 propane-triol). Examples for other completely miscible liquids are 1,4 dioxane and acetone (2-propanone), whereas diethyl ether is scarcely soluble in water.

The physical properties of the mixtures cannot be derived by means of the mixing rule and this shows that each liquid mixture is an "individuality". In particular we shall be concerned with the properties of water-alcohol mixtures in Chapter 12.

Hydrophilic Solutes

A large group of solutes in water is provided by hydrophilic solutes (see Chapter 11) which are dissolved due to the strongly developed coordinating properties of water. Many ionic compounds are readily soluble to give conducting solutions and the hydrated ions are freely mobile. Examples are provided by the halides of the alkali metals and the alkaline earth metals (except the fluorides of the latter, which are sparingly soluble).

A solution containing 357 g NaCl per litre can be readily prepared [1]. This solution has the approximate composition $NaCl.9H_2O$ and hence it may be considered as a molten hydrated salt.

Water is an excellent solvent for many acids and bases and hence for the performance of neutralization and hydrolysis reactions. Thermodynamic studies have led to the assessment of acidity and basicity constants as well as of thermodynamic data for neutralization and hydrolysis reactions, as well as for complex equilibria.

Hardly soluble hydrophilic solutes have small enthalpies of hydration and require for their stabilization a much greater number of specifically rearranged water molecules than solutes of high solubilities. For example, carbon dioxide is moderately soluble with formation of hydrated hydrogen ions and hydrated hydrogen carbonate ions. Addition of sugar or sodium chloride leads to "salting out" of carbon dioxide because of the higher enthalpies of hydration of sugar or sodium chloride. Water molecules used for the stabilization of the hydrogen carbonate ions are taken away from them. This shows that *the solution acts as a unity and all of the parts of the solution are affected by the dissolution process and by the presence of solutes.*

Although solid silver chloride has the same crystal structure as sodium chloride, it has a very low solubility and this is characterized by the *solubility product*. One of the most insoluble compounds is mercuric sulphide, found in nature as cinnabar and formed in the laboratory by precipitating a solution of a mercuric salt with hydrogen sulphide. According to its solubility product of 4.10^{-52} at 20°C 2.10^{-26} mol HgS are soluble in one litre of water, corresponding to the presence of one molecule mercuric sulphide in 333 litre. This means an extremely low probability to find one molecule in one millilitre of the solution.

This result need not alarm us, as knowledge of the solubility product allows the calculation of the solubility in the presence of other solutes. For example, in the presence of 10^{-6} mol mercuric chloride the solubility of mercuric sulphide is 4.10^{-46} mol HgS per litre. This means that the whole water content on earth would not be sufficient in order to produce a solution which definitely would contain one molecule of mercuric sulphide.

Confronted with this situation we are reminded of the comment made by *Lewis* and *Randall* [2] on the vapour pressure of tungsten metal at room temperature. The vapour pressure of this metal is estimated by extrapolation of the results of measurements at very high temperatures, and the most striking result of a vapour pressure of 10^{-105} atm is obtained for the temperature of +100°C. This would mean that the concentration of tungsten vapour would be less than one atom in a space equivalent to the known sidereal universe. Nevertheless, even this low value of estimated vapour pressure might be used in thermodynamic work with the same sense of security as we use the vapour pressure of water.

It is also well-known that many organic substances are appreciably soluble in water. Examples are most of the carbohydrates, amines, ketones, hydroxyl compounds or urea and its derivatives.

Hydrophobic Solutes

Whereas hydrophilic solutes owe their solubilities to the formation of hydration spheres, hydrophobic solutes do not form such hydration shells in water. Instead, the solute molecules are placed in polyhedra formed by water molecules and these constitute cages for the dissolved hydrophobic molecules [3] (Chapter 10).

Examples for hydrophobic solutes are the inert gases, nitrogen, oxygen or methane. Their solubilities are low, but it has not been possible to produce a liquid absolutely free from any dissolved gases [4]. This led us to the conclusion that the presence of gas molecules is one of the requirements for the existence of a liquid and this will be elaborated in more detail in Chapters 10 and 17.

Table 7 shows that the composition of air in the atmosphere is different from that in sea water.

Table 7. Composition of air in the atmosphere and in sea water.

Gas	In the atmosphere at 20°C [%]	ml gas per litre sea water at 12°C
N_2	78.08	11.1
O_2	20.95	6.2
Ar	0.93	0.3
CO_2	0.03	0.3

Other hydrophobic solutes are aliphatic and cyclic hydrocarbons, carbon tetrachloride and even ionic compounds with ions containing hydrophobic groups, such as tetrabutylammonium halides, or tetraphenylborates (Chapter 10).

Amphipathic Solutes

A special class is provided by the amphipathic solutes. These are molecules or ions which exhibit both hydrophilic and hydrophobic properties simultaneously [5]. The molecules are characterized by a "head-group" that is strongly hydrophilic coupled to a hydrophobic "tail", usually a hydrocarbon group.

Their association follows the similar rule, according to which like regions interact with each other, i.e. the hydrophilic head groups associate with each other just as well as the hydrophobic tails associate to form layers or zones bordered by layers of hydrophilic residues [5].

In this way lipid molecules tend to aggregate both in the dry as well as in the wet state. In the wet state the hydrophilic groups are anchored in the water structure, whereas the hydrophobic chains behave like tails of the molecules. Molecules with long tails may form a monolayer at the water surface with only the head group immersed in water. They decrease considerably the surface tension of the liquid and are known as surfactants (see Chapter 9).

Alternatively, if the mixture is vigorously stirred, *micelles* are formed. These provide spherical structures formed by a single layer of molecules [6] and these will be considered in more detail in Chapter 9.

Of particular importance for living systems is the formation of bilayer vesicles. In these the hydrocarbon tails of the molecules tend to lie in roughly parallel arrays tail to tail, so that bilayers are formed with the strongly hydrated hydrophilic groups forming an interface towards the water. They form the basis of biologically membranes, such as cell membranes. These provide partitions between cellular compartments, by which intracellular and extracellular water appear both separated and connected (see p. 91 ff and p. 209 ff).

3. Reactivity of Water

Self-Ionization

It has been mentioned that apart from the hydrogen fluoride molecules no other molecules are known which could compete with water molecules with regard to equally well-developed donor and acceptor functions. By the mutual interactions of the water molecules in the liquid state a three-dimensional network of great flexibility and of great strength is established.

Another manifestation of the well-developed donor and acceptor properties of the water molecules is the self-ionization equilibrium in liquid water. This appears to result from the cooperative interactions of water molecules. Due to the highly developed donor and acceptor functions of the water molecules, they are subject to continuously occurring cooperative interactions [4].

If the hypothetical dimeric unit of water molecules presented on p.44 is considered as embedded within the water structure and subject to cooperative interactions between all of the hydrogen bonded water molecules, the following situation may be considered:

Due to the concerted actions of the water molecules surrounding the dimeric unit (indicated in the preceding presentation by strong arrows) the intermolecular bond $O_{(1)}$ - $H_{(3)}$ is shortened to an extent that it obtains the characteristics of an intramolecular bond. The intramolecular bond $H_{(3)}$ - $O_{(2)}$ is lengthened to an extent that it is heterolyzed. The result is the formation of a hydrated H_3O^+ -ion and of a hydrated OH^- -ion, both within the water structure [7].

Similar considerations apply for the self-ionization equilibria in other hydrogen bonded solvents. The ionic product of liquid hydrogen fluoride is greater than that of water, because the hydrogen bond energy is greater in the former and the cooperative effects acting in one-dimension only. The equilibrium constant is smaller for liquid ammonia because of the weaker hydrogen bonding in this solvent.

Table 8. Self-ionization equilibria and autoprotolysis constants

Equilibrium			Ionic Product	pH of neutral Solution
2 HCOOH	$\rightleftharpoons$ $[HCOOH_2]^+$	+ $[HCOO]^-$	10^{-6}	3
2 HF	$\rightleftharpoons$ $[H_2F]^+$	+ F^-	10^{-10}	5
2 H$_2$O	$\rightleftharpoons$ $[H_3O]^+$	+ OH^-	10^{-14}	7
2 C$_2$H$_5$OH	$\rightleftharpoons$ $[C_2H_5OH_2]^+$	+ $[C_2H_5O]^-$	10^{-20}	10
2 NH$_3$	$\rightleftharpoons$ $[NH_4]^+$	+ NH_2^-	10^{-28}	14

Hydrolysis Reactions

Hydrolysis reactions are another manifestation of the high reactivity of liquid water. Many covalent compounds are readily hydrolysed, such as silicon tetrachloride. Even ionic compounds may be hydrolysed, such as phosphorus pentabromide, which crystallises in the solid state in a caesium chloride lattice [8] or calcium carbide which is hydrolysed to acetylene and calcium hydroxide.

As the materials exposed to water are decomposed by hydrolysis, the reaction products may be soluble in water or provided in such highly dispersed states that they remain in the aqueous solution.

Redox Reactions

Several metals are known to react violently with liquid water. For example, sodium gives off hydrogen with formation of sodium hydroxide. According to the electromotive series, metals which are less noble than iron should be attacked by water. Such corrosion processes are known for iron and many of its alloys, whereas other less noble metals, such as magnesium or aluminium are known to be passivated by air.

Potable water is frequently conveyed through lead pipes. While lead is readily attacked by distilled water, it is resistant towards tap water, which contains enough oxygen and carbonates to provide an insoluble layer on the surface of the metal. Hard water is known to have less action on lead than soft water.

Noble metals, such as gold or silver, do not displace hydrogen from water. It is, however, known that small grains of these metals are suspended in river water and in this form transported to the sea.

It is also possible to prepare solutions containing colloidal gold. They are formed by reducing gold chloride with hydrazine or formaldehyde. Solutions of larger particles are blue and of small particles ruby-red (silver solutions are yellow and copper solutions red). By reducing gold chloride with tin(II)-chloride, a purple colloid is formed. The purple powder obtained from this colloidal solution is called purple of *Cassius* and may be used for making ruby glass by fusing the glass with this powder followed by annealing.

The actions of small amounts of solutes on the quality of water are well-illustrated by the enormous variety of *mineral waters* found in nature.

4. The Hydrologic Cycle

The path taken by water from the oceans to the atmosphere by evaporation, from the atmosphere to the ground by precipitation and ultimately back to the sea by river runoff, is known as the hydrologic cycle. This cycle provides for a relatively stable distribution of water between land, ocean and atmosphere (Table 9).

Table 9. Water balance on earth per year.

	Land [cu km]	World Ocean [cu km]
precipitation	+ 108.000	+ 412.000
river runoff	- 37.000	+ 37.000
evaporation	- 71.000	- 449.000

The total balance of free water has probably been constant for 500 million years. Evaporated water is precipitated manly as rain which contains dissolved gases and suspended particles from the atmosphere. As it runs down mountains and hills it gradually picks up nutrients and makes them available to regions below. It penetrates soil, the water of which is used by plants. It erodes and sculpts the rocky surface of the earth and transports nutrients, inorganic solutes and suspended sediments into swamps, lakes and rivers and eventually the ocean.

The rate of water circulation through the rain-river-ocean-atmosphere system or hydrologic cycle is relatively fast. The amount of water discharged into the oceans each year from land is about 37.000 cu km (Table 9) and this is approximately equal to the total mass of water stored at any instant in rivers, lakes and ground water.

River water which has previously percolated through soil contains many solutes and suspended matter, both mineral (clay) and organic. Solutes may be ammonium salts, nitrites, sodium chloride etc. as well as organic matter of vegetable and animal origin. It contains dissolved atmospheric gases, the oxygen of which is of importance for fish. Oxygen is derived from the atmosphere and from marine plants that use sunlight to produce carbon dioxide by photosynthesis, but oxygen is consumed everywhere by organisms and by combustion of organic waste products when the dead remains of organisms sink to the bottom. Because of the density maximum of water at +4°C, water cooled to this temperature sinks and reaches, charged with oxygen, the ocean floor, where it spreads out by movement of ground water as part of the hydrolytic cycle.

Sea water contains nearly all metal ions at least in trace amounts as well as small grains of noble metals, such as gold and silver.

Table 10. Distribution of water on earth.

Water Resources	Water Volume [cu km]	%
Entire Earth	1 430 000 000	
Oceans	1 370 000 000	96.0
Earth Crust	60 000 000	4.0
including		
glaciers	29 000 000	2.0
lakes	750 000	0.05
soil	65 000	0.005
atmosphere	14 000	0.0008
rivers	1 200	0.00008

Because our experience is hardly accustomed to the dimensions both on a world wide scale and on a molecular scale, the following considerations may be made:

1. The number of sodium and chloride ions in each litre of sea water is about 3.10^{23}. If all of the salts contained in the oceans would be extracted, all land on earth could be covered by a salt layer about 100 m high.

2. If the gold content of the oceans (Table 11) could be extracted, a cube of pure gold metal may be obtained with edge length of 70 m.

3. Suppose that all of the water molecules contained in one litre of water (3.10^{25} water molecules) are labelled, poured into the sea and even distribution over the whole water content assumed to take place, one would find in each litre of water 25 000 labelled water molecules.

By means of rivers about 420 000 000 tons of constituents are annually carried to the oceans which receive in this way much of industrial, municipal and agricultural waste products. The lead content of surface sea water is increasing rapidly and oil appears to be polluting the surface water on a world wide scale.

Within the hydrologic cycle, the accumulation of many elements by organisms is extremely important. Carbon is dramatically enriched by all organisms. Small concentrations of many solutes in sea water are substantially increased by the abilities of marine invertebrates to accumulate for their own use various trace elements from sea water, but also chemically persistent elements are accumulated by them.

For example, DDT, carried to the sea, is accumulated in the fatty tissues of organisms and becomes concentrated as these are transferred into the food chain.

Many marine invertebrates have mineralized skeletons, in which various elements have been concentrated from sea water.

Table 11. Elements in Sea Water [9].

        The elements in sea water are listed in decreasing average concentration and they contain also the elements of water (oxygen and hydrogen). These are average concentrations and there will be variations in composition as a function of the location where the sample was collected.

Element	Concentration [mg/l]	Element	Concentration [mg/l]
Oxygen	$8.57 \cdot 10^5$	Vanadium	$2 \cdot 10^{-3}$
Hydrogen	$1.08 \cdot 10^5$	Titanium	$1 \cdot 10^{-3}$
Chlorine	$1.90 \cdot 10^4$	Cesium	$5 \cdot 10^{-4}$
Sodium	$1.05 \cdot 10^3$	Cerium	$4 \cdot 10^{-4}$
Magnesium	$1.35 \cdot 10^3$	Antimony	$3.3 \cdot 10^{-4}$
Sulfur	$8.85 \cdot 10^2$	Silver	$3 \cdot 10^{-4}$
Calcium	$4.00 \cdot 10^2$	Yttrium	$3 \cdot 10^{-4}$
Potassium	$3.80 \cdot 10^2$	Cobalt	$2.7 \cdot 10^{-4}$
Bromine	$6.5 \cdot 10^1$	Neon	$1.4 \cdot 10^{-4}$
Carbon	$2.8 \cdot 10^1$	Cadmium	$1.1 \cdot 10^{-4}$
Strontium	8.1	Tungsten	$1 \cdot 10^{-4}$
Boron	4.6	Selenium	$9 \cdot 10^{-5}$
Silicon	3	Germanium	$7 \cdot 10^{-5}$
Fluorine	1.3	Xenon	$5.2 \cdot 10^{-5}$
Argon	$6 \cdot 10^{-1}$	Chromium	$5 \cdot 10^{-5}$
Nitrogen	$5 \cdot 10^{-1}$	Thorium	$5 \cdot 10^{-5}$
Lithium	$1.8 \cdot 10^{-1}$	Gallium	$3 \cdot 10^{-5}$
Rubidium	$1.2 \cdot 10^{-1}$	Mercury	$3 \cdot 10^{-5}$
Phosphorus	$7 \cdot 10^{-2}$	Lead	$3 \cdot 10^{-5}$
Iodine	$6 \cdot 10^{-2}$	Zirconium	$2.2 \cdot 10^{-5}$
Barium	$3 \cdot 10^{-2}$	Bismuth	$1.7 \cdot 10^{-5}$
Aluminium	$1 \cdot 10^{-2}$	Lanthanum	$1.2 \cdot 10^{-5}$
Iron	$1 \cdot 10^{-2}$	Gold	$1.1 \cdot 10^{-5}$
Indium	$2 \cdot 10^{-2}$	Niobium	$1 \cdot 10^{-5}$
Molybdenum	$1 \cdot 10^{-2}$	Thallium	$1 \cdot 10^{-5}$
Zinc	$1 \cdot 10^{-2}$	Hafnium	$8 \cdot 10^{-6}$
Nickel	$5.4 \cdot 10^{-2}$	Helium	$6.9 \cdot 10^{-6}$
Arsenic	$3 \cdot 10^{-3}$	Tantalum	$2.5 \cdot 10^{-6}$
Copper	$3 \cdot 10^{-3}$	Beryllium	$6 \cdot 10^{-7}$
Tin	$3 \cdot 10^{-3}$	Protactinium	$2 \cdot 10^{-9}$
Uranium	$3 \cdot 10^{-3}$	Radium	$6 \cdot 10^{-11}$
Krypton	$2.5 \cdot 10^{-3}$	Radon	$6 \cdot 10^{-16}$
Manganese	$2 \cdot 10^{-3}$		

This shows again the unique and unavoidable role of water in all cycles in nature, such as metabolism. This embraces all of the chemical changes that occur in the cells of living organisms, enabling them to grow, to maintain their identity and to reproduce.

Table 12. Concentration factor for some elements in marine organisms with respect to sea water.

Element	Conc. Factor	Element	Conc. Factor
V	280.000	Mo	6.000
I	200.000	As	3.300
Fe	86.000	Pb	2.600
Mn	41.000	Sn	2.700
Ni	41.000	Cr	1.400
Zn	32.000	Au	1.400
Ag	22.000	Sr	1.200
Co	21.000	Bi	1.000
Ti	10.000	Ga	800
Ge	7.600	Tl	700
Cu	7.500	Sb	300

Some marine invertebrates consist to 97 % of water and the human body contains 65 to 70 % water, which is unevenly distributed within the organism.

Table 13. Water content within the human body.

Intracellular Water	45 - 50 %
Extracellular Water	15 - 20 %
Plasma Water	5 %
Regions of small Water Content	30 %
Nervous Tissues	84 %
Liver	73 %
Muscles	77 %
Skin	71 %
Connective Tissues	60 %

References

1. *Handbook of Chemistry and Physics*, 73rd edn. (CRC Press, Cleveland/Ohio 1992).
2. G. N. Lewis and M. Randall, *Thermodynamics*, Revised by K.S. Pitzer and L. Brewer, 2nd edn. (McGraw Hill, Series in Advanced Chemistry, New York, 1961).

3. C. Tanford, *The Hydrophobic Effect: Formation of Micelles and Biological Membranes* (Krieger Publ. Co. Malabar/Florida, USA, 1991).
4. G. Resch and V. Gutmann, in *Advances in Solution Chemistry*, eds. I. Bertini et al, p.1, (Plenum Press, New York, 1981).
5. C. K. Matthews and K. E. van Holde, *Biochemistry*, (Benjamin/Cunnings Publ. Co., Redwood City/Calif., 1990).
6. H. Hausner, in *Water a Comprehensive Treatise*, ed. F. Franks, Vol. 4, Chapter 4, p. 209 (Plenum Press, New York, 1975).
7. V. Gutmann and G. Resch, *Z. Chem.* **19** (1979) 406.
8. M. van Driel and C. H. McGillavry, *Rec. Trav. Chim.* **62** (1943) 167.
9. A. Demayo, in *Handbook of Chemistry and Physics*, ed. D.R.Lide, 73rd edn. CRC Press, Cleveland/Ohio, 1992).

CHAPTER 8

THE PHASE BOUNDARY OF LIQUID WATER

1. Introduction

In order to obtain insight into the differentiations of aqueous solutions, we may, first of all, draw our attention to the molecules at the phase boundaries. It may be trivial to emphasize

(i) that a phase boundary is necessary for the observation and for the very existence of the liquid and

(ii) that all processes in living systems involve chemical and physical interactions of water at and near the phase boundaries [1].

The practical importance of the effects at the phase boundaries of water are also obvious, for example, with regard to lubrication, flotation, foaming, adhesion, emulsification, wetting or corrosion phenomena.

2. Surface Tension

It appears as if by means of the phase boundaries the liquid is held together by a kind of skin or membrane. This implies that the water molecules arranged at the phase boundary must have "unexpected" properties. They are in states of higher energy for which the following description is provided on the molecular level.

A molecule in the interior of liquid water has a similar environment in all directions, but a water molecule at the interface is under a resultant attraction inwards because the number of water molecules per unit volume is greater in the bulk of the liquid than in the vapour. A molecule at the interface is subject to an inward-pull, so that the surface of a liquid always tends to contract to the smallest possible area and so that drops of a liquid and bubbles of a gas within a liquid become nearly spherical.

In order to extend the interface area it is necessary to do work namely in order to bring molecules from the bulk to the interface. The work to increase the interface area by 1 cm^2 is called the *free surface energy*. The molecules are in a state of tension and this is known as the *surface tension*, which can be measured.

The surface tension of liquid water of is much higher than that of most other liquids and it is decreased as the temperature is increased. The surface tension becomes zero and the densities of the liquid and of the gas reach the same values, when the critical point is reached (for water at +374°C and 194.6 atm pressure). This means that under these conditions the state of tension can no longer be maintained and that in its absence the liquid cannot exist any longer. The state of tension is therefore required for the functions of a phase boundary and for the existence of a liquid and not a "fault" in its structure.

3. Characterization of a Phase Boundary

The state of tension is the result of different properties of the phases forming the phase boundary. Because of their mutual interactions, the phase boundary of a given liquid is at the same time the phase boundary of the adjacent phase, which may be solid, liquid or gas. The two phases are separated from each other by the discontinuity of the phase boundary and at the same time interconnected with each other. It is therefore also called an *interface*.

Because of the mutual interactions at an interface, the value measured for the surface tension is always influenced by the properties of the adjacent phase. For this reason, it is not possible to obtain highly precise and strictly reproducible values for the surface tension [2].

For macroscopic observation the phase boundary provides a discontinuity and hence the two phases appear clearly separated from each other.

For the microscopic observation the border line between the two phases is less clear-cut. Due to the molecular interactions, such as adsorption, desorption, evaporation or deposition of particles, local inhomogeneties are always present. These concern local and temporal differences in morphological, analytical, structural and energetic features and it is not possible to clearly distinguish to which of the two phases the individual particles belong.

The continuous changes imply also that the molecules at the interface cannot be in thermodynamic equilibrium, but rather maintain dynamically the conditions as they are required for the maintenance of the equilibrium of the whole liquid under the given environmental conditions.

If it were possible to consider the dynamic aspects in detail, the observer would be so confused by the continuous interactions that he would find himself unable to draw a border line at all.

4. The Electrical Double Layer

In order to account for the state of tension at the electrode-electrolyte interface the concept of the electrical double layer has been advanced. Thus the electrical double layer is considered to be the result of the presence of ions, which - in the early theories - are treated as point charges of negligible volume within the dispersion medium. The latter is considered to influence the double layer through its permittivity. All advancements in theory are also based on the presence of ions firmly held together within a "diffuse" region [1].

The electrical double layer is considered as providing the potential difference, which is established at the interface of a metal in a solution of its ions. This system is called a *half-element*, but the potential difference at its interface cannot be measured. Measurable is, however, the difference of the potential differences of two half-elements. If exactly the same half-elements are shunted together, no potential difference is found, but if the two half-elements differ from each other (see p. 201), for example, in the electrolyte concentration, a potential difference can be measured. This arrangement is also known as a *concentration cell*.

When two solutions each of them with a different concentration of the same electrolyte are brought in contact, the more concentrated solution will tend to diffuse into the more dilute. The driving force for this process is the presence of a state of tension at the junction of the two solutions and this is known as the *liquid junction potential*.

A liquid junction potential is found at each liquid junction, such as between solutions of different concentrations. If iodine solutions of different concentration in carbon tetrachloride come into contact, the liquid junction potential obviously cannot be accounted for by the application of an ionic model.

An interesting result has been obtained by *Horne et al* [3] who passed a solution of electrolytes through a column filled with solid particles and compared the electrical conductivity of the eluted solution with that of the original solution (the solid particles are permeated with the solution). They concluded that water near an interface tends to "exclude" electrolytes, thereby increasing the concentration of the solution at some greater distance from the interface.

It has been shown in Chapter 4 that physical and chemical interactions can be described in terms of the extended donor-acceptor concept [4]. Because of the continuous interactions between the phases at the interface, pileup effects of negative charge occur at the donor sites and spillover effects of negative charge at the acceptor sites. In these ways partial electrical charges are established and dynamically maintained at each interface as long as a phase boundary is present. This electrical double-layer and the state of tension provide appropriate reactivities at the interfacial region and the "openness" of the liquid to its environment through its phase boundary.

5. Bond Length Considerations

According to the bond length variation rules a smaller coordination number of the interface molecules means stronger bonding and hence a bond contraction at the interface. Because by interactions with the environment bond lengthening is expected to occur and because this will depend on the interaction energy, different bond lengths are expected in different environments.

Unfortunately, there is no experimental method available to measure bond distances at the interface of a liquid. It may therefore be convenient to provide a short description about the state of affairs at phase boundaries of solid materials. Based on quantum mechanical considerations *Lennard-Jones* [5] predicted in 1928 that lattice parameters near and perpendicular to the surface should be smaller than within the crystal lattice. The first experimental indication for this prediction was obtained 23 years later by *Boswell* [6] and later confirmed by the results of modern spectroscopic techniques, namely the low energy electron diffraction (LEED) method. The results of these investigations have been reviewed [7].

It has further been shown that the O-H bonds at the surface of a given silica material are found longer the stronger the interactions of the terminal Si-O-H bonds due to adsorption by materials attacking the hydrogen atoms [8]. Even adsorption of benzene leads to a lengthening of the O-H bond. *Horill* and *Noller* [8] followed the changes in O-H-valence frequency accompanying adsorption of a variety of donor molecules on aerosol and found that the values of the frequency shifts of the O-H bond varied linearly with the donor number of the adsorbate.

$$—Si—O—H \leftarrow Donor$$

increased
bond length

6. Water in Thin Layers

The interfacial region of liquid water is highly differentiated and cannot be considered as a simple or sharp termination of the bulk structure. It is, however, extremely difficult to investigate these interface effects experimentally, unless thin layers of water are studied. This is because the number of interface molecules is under other circumstances too small and because all local effects are disguised in the statistical data.

The thermal conductivity of layers of thickness of about 1µm between mica sheets was found by one magnitude greater than that of bulk water and about three times that of ice [9].

Another remarkable effect has been reported by *Anderson* and *Low* [10]. They investigated the density of water as a function of the water content for a sodium bentonite at 25°C and observed a decrease in density over a distance of 6nm from the clay surface. At about 1nm from the surface the density reduction amounts to 2.5% of the bulk water density at 25°C [10] (Fig. 16).

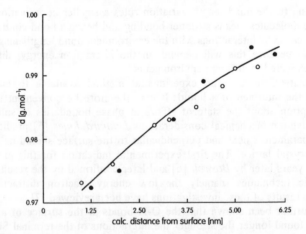

Fig. 16. Density of water adsorbed on sodium bentonite at 25°C as a function of the thickness of the water layer obtained in two different experimental runs [10].

Similar results were obtained on water in 14 nm diameter silica pores, which was found by 3% less dense than bulk water [11]. Such volume changes have also been obtained by means of high-precision density measurements of suspensions of solid particles [12]. The water layer near the solid was found less dense, but it was not possible to derive the thickness of the affected zone from such measurements, since only overall effects could be observed. It was calculated, however, that for water the effect was compatible with a zone 3 nm thick in which the density is 1% less than in the bulk liquid.

Another interesting result is the higher heat capacity of water in thin layers. The heat capacity of liquid water in 24 nm diameter silica bores was found to be about 10% greater than in bulk water and the minimum in heat capacity at a lower temperature, namely about 24°C instead of 37.5°C in bulk water (Fig. 17)[13].

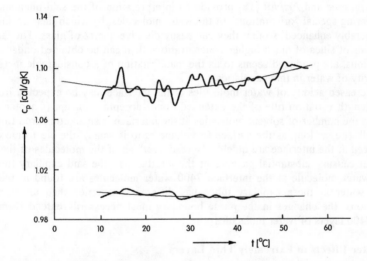

Fig. 17. Heat capacity of bulk water (lower curve) and of water in 24 nm silica pores (upper curve) as a function of temperature [13].

Another interesting result concerns the surface tension of water condensed in a micropore of 20 nm radius, which is much lower than that of bulk water.

The properties of clay minerals used in ceramics [14,15] and in the brick and cement industries are essentially dependent on the clay-water interactions for the achievement of optimal plasticity and coherence.

Clays, mainly kaolins, are used by the paper industry as filler materials and coatings. As in ceramics, the guiding knowledge has been mainly empirically acquired and leaves plenty of room for fundamental research, in particular on questions related to the water balance between the cellulose network and the coating.

Zeolites are known for their ion-exchange and reversible dehydration properties. Their structures provide channels of various diameters, which admit water molecules and certain ions. They are used as water softeners and as molecular sieves.

Chemical effects ranging over layers of even greater thickness have been derived from the results of investigations of the so-called anomalous water. A highly concentrated solution containing mainly silicic acid is formed, when water is condensed in capillary tubes of silica 2 to 4 μm in diameter. This solution has been reported by *Deryagin* [16] and interpreted as a new modification of water, called "polywater". With regard to the last mentioned interpretation, we are perfectly justified in talking about the "legend of polywater". On the other hand it is a fact and not a legend that water is unexpectedly reactive in extremely thin layers and capable of dissolving silica. This is passed into the aqueous phase in concentrations which are much higher than expected from solubility data [17].

Prigogine and *Fripiat* [18] provide an interpretation of the said phenomena by considering special "orientations" of the water molecules, by which their acidities are considerably enhanced, so that they can easily dissolve quartz of glass. The fact that solutions of silica of much higher concentrations than can be obtained under normal conditions, are produced seems to us the manifestation of a considerable increase in reactivity of water in thin capillaries.

Increased acidity of water molecules at the interface may be expected from the bond length variation rule of the extended donor-acceptor concept (Chapter 4). As long as the number of solvent molecules at the interface is small compared to that in the bulk (i.e. as long as the surface to volume ratio is small), the electronic effects produced at the interface are quickly "spread over" all of the molecules of the liquid without causing substantial changes at the interface. In the said capillary tubes for each water molecule at the interface 7400 water molecules are found in the bulk. Since water in these capillary tubes has different properties than under normal conditions, the changes in the phase boundary must necessarily extend themselves over 7400 layers of water molecules.

7. Water Effects in Extremely Thin Layers

Interesting is the fact that water in extremely thin layers exhibits very low phase transition temperatures [19]. Fig.18 shows that a "bilayer" of water molecules appears to remain liquid down to a temperature of -38°C, i.e. it behaves like supercooled water (Chapter 6) and hence the water molecules in the thin layer are dynamically considerably more active than in the bulk.

A theoretical approach led to the conclusion that at liquid-vapour interfaces long-range ordering is established within the interfacial zone. The physical manifestation of this order are described as *surface waves*, characteristic interfacial modes in which the motions of widely separated portions of the transition zone are strongly correlated [20].

Unfortunately it is extremely difficult to learn about thin water layers from modern spectroscopic techniques, because the individual features in the border line areas are disguised within the statistical data. As a distinction can be made between ice and liquid water, these techniques revealed that in thin layers some water remains unfrozen down to very low temperatures.

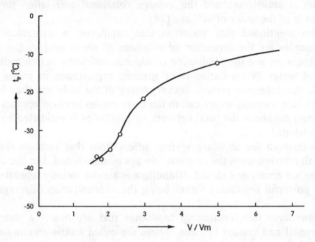

Fig. 18. Phase transition temperature t_{tr} plotted against V/V_m, V_m representing a monolayer capacity and V the adsorbed amount of water [19].

8. Water at Interfaces in Biological Systems

Similar results were obtained for water around biological macromolecules, tissues, and DNA molecules[21]. The infrared adsorption spectra revealed that an inner layer of about ten water molecules per nucleotide is not freezing at -60°C and hence retaining considerable motional freedom.

This shows the enormous role of fluidity of water layers around the DNA molecules, which loose their genetic information by the process of dehydration. The genetic information is not regained by re-hydration and this shows the enormous role of water.

All biopolymers (proteins, nucleic acids, peptides) are hydrated to a considerable degree. Since, as mentioned before, at least some of this hydration water does not freeze at low temperatures, it must have been "dynamized" by its environment. It seems as if the DNA structures provide conservative boundary conditions for the genetic information which is dynamically maintained by the continuous interactions with and within the aqueous phase, in which the structures have been produced.

More interest has been devoted to the role played by water in stabilizing certain conformations of the macromolecules in preference to many other "statistically available" conformations [22].

Trincher [23] has suggested that water in contact with the erythrocyte surface has "protective properties by virtue of its structure" and that the water adjacent to the cell surface may exist in two different microphases with transitions occurring between them at different temperatures.

Studies on fibrous collagen and on DNA have shown that the reorientation of water molecules is anisotropic and the average rotational correlation time of the water molecules is of the order of 10^{-7} sec.[24].

It has been mentioned that water in thin capillaries is dynamically highly developed. Examples are the formation of solutions of silicic acid by the action of water in thin capillaries and the production of supercooled water in thin capillaries.

The role of water in capillaries is of greatest importance in living systems. Microscopically thin tubes are present in every tissue of the body and they form such a dense network that virtually every cell in the body comes in close contact with the blood. In a human organism the total network of capillaries is estimated to be some 60 000 miles in length!

Bloodflow through the capillary system differs from that through the arterial system in that all arteries, even the smallest, always contain blood, but the capillaries most of the time are empty and closed. Bloodflow is almost entirely regulated locally with the most powerful regulating factor being the concentration of oxygen in the given tissue.

Some organs have non-functioning capillaries that are nearly a thoroughfare between the arterial and venous systems. These are called arteriovenous capillaries. They are particularly numerous in the skin, where they play an important role in the body's heat regulation.

The principal functions of the blood is the supply of all cells with the material "they need" and the removal of the waste products from the cells. This means that *the surface structures are dominated by water* and that the information exchange between different aqueous regions occurs via interfaces.

The porous walls of the capillaries act as permeable membranes with selective permeability for different substances with highest permeability for water [1]. Active exchange of water between plasma and tissue fluid occurs on a basis other than diffusion. It is driven into and out of the blood by pressure. There is a constant exchange of water between blood plasma and extracellular fluid with the water leaving the capillary at its arterial end and returning to the capillary at its venous end [25].

The magnitude of this capillary exchange can be appreciated by knowing that three litre of blood plasma coming in contact with fifteen litre extracellular fluid, exchange the blood volume some 45 times every minute.

The enormous role of amphipathic solutes for the formation of interface layers, such as cell membranes and for their functionalities will be briefly outlined in the following Chapter as well as in Chapter 19.

References

1. J. Clifford, in *Water, a Comprehensive Treatise*, ed. F. Franks, Vol. 5, Chapter 2, p. 75, (Plenum Press, New York, 1965).
2. H. Sobol, J. Garfias and J. Keller, *J. phys. Chem.* **80** (1976) 1941.
3. R. A. Horne, A. F. Day, R. P. Young and N. T. Yu, *Electrochim. Acta* **13** (1968) 397.
4. V. Gutmann, *The Donor-Acceptor Approach to Molecular Interactions*, (Plenum Press, New York, 1977).
5. E. J. Lennard-Jones, *Proc. Roy. Soc*, **A 121** (1928) 247.
6. F. W. C. Boswell, *Proc. Phys. Soc.* **A 64** (1951) 465.
7. V. Gutmann and H. Mayer, *Rev. Inorg. Chem.* **1** (1979) 51.
8. W. Horill and H. Noller, *Z. physik. Chem. (n.F.)* **100** (1976) 155.
9. M. S. Metsik and O. S. Aidanova, in *Research in the Field of Surface Forces*, ed. B.V.Deryagin, Vol.2, p.169 (Consultants Bureau, New York, 1966).
10. D. M. Anderson and P. L. Low, *Soil Sci. Soc. Amer. Proc.* **22** (1958) 99.
11. F. M. Etzler and D. M. Fagundus, *J. Coll. Interface Chem.* **93** (1983) 585 - **115** (1987) 513.
12. S. G. Ash and G. H. Findenegg, *Trans. Farad. Soc.* **67** (1971) 2122.
13. F. M. Etzler and J. J. Conners, *Langmuir* **6** (1990) 1250.
14. E. Forslind and A. Jacobsson, in *Water, a Comprehensive Treatise*, ed. F. Franks, Vol. 5, Chapter 4, p. 173, (Plenum Press, New York, 1975).
15. R. E. Grim, *Applied Clay Mineralogy*, (McGraw Hill, London, 1962).
16. B. V. Deryagin, *Sci. Am.* **223** (1970) 52, 69.
17. D. Schüller, *Naturwiss.* **60** (1973) 145.
18. I. Prigogine and J. J. Fripiat, *Chem. Phys. Lett.*, (1971) 107 - *Bull. Soc. Chim. (France)* (1971) 4291.
19. M. Chikazawa, T. Kanazawa and T. Yamaguchi, *Kona Powder Sci. Techn. Japan* (1984) 54 - A. Tsugita, T. Takei, M. Chikazawa and T. Kanazawa, *Langmuir* **6** (1990) 1461.
20. M. S. Shon, R. C. Desai and J. S. Dahler, *J. chem. Phys.* **68** (1978) 5615.
21. M. Falk, A. G. Poole and C. G. Goymour, *Can. J. Chem*, **48** (1970) 1536.
22. D. Eagland, in *Water, a Comprehensive Treatise*, ed. F. Franks, Vol. 4, p. 305, (Plenum Press, New York, 1975).
23. K. Trincher, *Die Gesetze der biologischen Thermodynamik*, (Urban and Schwarzenberg, Wien - München, Baltimore, 1981).
24. G. Migchelsen, H. J. C. Berendsen and A. Rupprecht, *J. Mol. Biol.* **37** (1968) 235.
25. L. P. Kayashin, *Water in Biological Systems* (Consulatants, Berlin, New York, 1969).

CHAPTER 9

WATER IN BIOLOGICAL SYSTEMS

1. Amphipathic Solutes

The presence of amphipathic solutes is of enormous importance for the unique role of aqueous solutions in biological systems. The term *amphipathic* is of Greek origin (amphi = dual, pathi = sympathy) and is applied to molecules which contain both a hydrophilic group and one or two long hydrophobic chains.

Amphipathic solutes tend to aggregate both in the dry state and in water. Association of long-chain lipids follow the *simile rule*, according to which like regions interact with each other: hydrophilic groups, which are also called "head groups" of the amphipathic molecules, associate with each other, just as well as hydrophobic residues, which are also called "tails" of the amphipathic molecules, associate to form hydrocarbon layers. Water is attracted by hydrophilic groups, but repelled by hydrophobic chains [1].

Aggregations of amphipathic solutes are known to occur in three different ways, namely with formation of
1. Surface layers,
2. Micelles,
3. Bilayers.

2. Surfactants

Because water is repelled by the hydrophobic chains, amphipathic solutes tend to reach the interface and to form monomolecular layers of well-aligned molecules. Their hydrophilic "head groups" are well anchored in contact with water, whereas the hydrophobic "tails" project out of the water thereby associating with each other at the interface with formation of a hydrophobic interface layer.

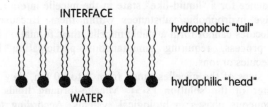

Fig. 19. Illustration of the arrangement of surface-active molecules at the interface.

This leads consequently to a marked decrease in surface tension of the liquid. For this reason, the molecules aggregating in this way are considered as surface-active agents or surfactants.

The study of surface monolayers has been known for a long time. For example, soap is one of the well-known surfactants and used for cleansing and other purposes. In view of the importance as wetting agents, emulsifiers and detergents a great number of surface-active substances has been prepared synthetically.

When all surface positions have been taken by amphipathic solutes and in particular, when the system is disturbed by shaking, micelles of high molecular weight are formed at the same time.

3. Micelles

Micelles are characterized by the strong association of hydrophobic tails with each other so that the arrangement of the hydrophilic head groups is directed towards water in such ways, that sphere-like aggregates are formed. In these the hydrophobic tails are completely shielded by the hydrophilic groups from the aqueous phase [2].

Fig. 20 illustrates that micelles contain a core of hydrophobic hydrocarbon groups surrounded by hydrophilic groups, which allow their existence in water.

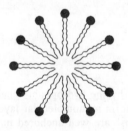

Fig. 20. Illustration of the molecular arrangement in micelles.

There is evidence for a "liquid-like" state in the micelle interior, because of the ability to dissolve hydrophobic substances within them. Because two or three amphipathic molecules cannot form a stable micelle, their formation is considered as a cooperative process, requiring simultaneous participation by water and amphipathic molecules or ions.

Because the sizes of the micelles range from 5 to 20 nm, they give rise to the colloidal character of the solution [3-5]. Micelle-forming lipids are of special importance in aqueous phases in biological systems. According to their melting points [1], two groups of lipids are distinguished:

(i) uncharged lipids, melting points rarely exceeding +70°C,

(ii) lipids containing charged groups, melting points between 200°C and 300°C.

Of particular interest is the phenomenon of polymorphism with long chain lipids. The prediction of such mesomorphic phases on grounds of chemical composition is not possible. The transition temperature T_c for the transition gel $\rightarrow$ liquid crystal depends on the water content and the presence of ions. By the said transition the molecular motion is increased and the packing density is decreased. The entropy increase observed in the course of this transition is characteristic for phase transitions in the solid state.

The lipid concentration at which micelle formation first occurs is known as CMC (Critical Micelle Concentration). This is altered by additives to the aqueous phase. It is interesting to note that for a given lipid the CMC in water is much lower than in alcohols. However, when a small amount of alcohol is added to water, the CMC is somewhat increased and after passing a maximum drastically decreased with increasing alcohol content.

One of the unsolved questions concerns the actions of urea or guanidinium chloride on the aggregation behaviour of amphipathic solutes. Urea is soluble in water, but after addition to a micelle-forming solution the CMC is increased. This increase in solubility of amphipathic solutes has been described as the "salting in of hydrophobic groups", including hydrocarbons [6]. In many cases, the presence of urea either inhibits the formation of inorganic gels, or at least reduces their mechanical strength considerably. The hydrophobic groups have a lower free energy in aqueous solutions of urea or guanidinium chloride than in water alone [7].

Tanford [8] comments: "Guanidinium chloride behaves like other electrolytes towards peptide groups, but clearly acts quite unlike inorganic salts towards hydrophobic groups. The reasons for this difference are not understood at present times... The answer to the question cannot come from studies of the effects of guanidinium chloride on the properties of water. It requires an investigation of the *molecular organization* in a three-component system hydrocarbon, guanidinium chloride and water."

4. Phospholipid - Bilayers

Instead of forming more or less sphere-like micelles, hydrophobic intermolecular interactions can also lead to the formation of so-called *bilayers*. These are formed by tail to tail interactions of the hydrophobic chains between two different layers, so that water-insoluble films are spread out within the liquid phase.

These films are spontaneously produced within the liquid structures and they remain parts of the liquid solution characterized by highly differentiated structures. *They separate different aqueous regions from each other and connect them at the same time.*

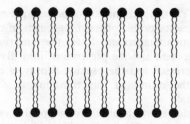

Fig. 21. Illustration of a phospholipid bilayer.

Phospholipid bilayers provide the basis for all biological membranes and as parts of the water structure they show high permeabilities for water.

The cell membrane provides boundary conditions, by which the highly specific interactions between extracellular water and intracellular water are taking place. The membranes have a thickness between 6 and 10 nm and these are part of the extracellular and the intracellular water, separated and connected by a double layer of hydrophobic lipids, mainly phospholipids and cholesterine.

The cells of higher organisms (eukaryotic cells) usually contain intracellular membranes that define compartments within the cell. Examples are the cell nucleus and the mitochondria, each being surrounded by its own membrane. Eukaryotic cells also contain endoplasmic reticulum, which is a network of membrane-bounded channels that traverse the entire cell. An exception is provided by the human erythrocyte, which contains no nucleus or other intracellular structures (see Chapter 4).

Proteins are integrated within the lipid bilayer and these are called integral membrane proteins. They are arranged at various positions in the range of the hydrophobic interactions. Intrinsic membrane proteins are inserted into membranes whereas extrinsic proteins interact with membranes primarily with the hydrophilic head group of the membrane (Fig. 22).

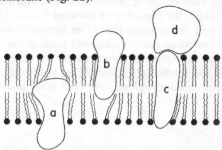

Fig. 22. Illustration of integral membrane proteins; a, b and c intrinsic membrane proteins, d extrinsic protein.

These proteins are currently the subject of intensive biochemical research [8]. As part of the membrane they are actually part of the water system, which seems to be decisive for the highly specific functions of the membrane proteins.

As water molecules accumulate between the hydrophilic head groups and the protein molecules of the phospholipid bilayer, a lateral expansion of the bilayer occurs and the area per lipid molecule is increased. The increased interactions at the head groups are paralleled by a decrease in interaction energy between the hydrocarbon chains and by a decrease in the density of packing inside the layer. This manifests itself in a decrease of the transition temperature for the gel → liquid crystal transition and in an increase in permeability of the membrane.

Consequently, the molecular packing of phospholipid molecules in the membrane bilayer is tightened and the transition temperature increased as the water content between the head groups is decreased.

If ions such as Na^+, K^+, Ca^{++}, Mg^{++}, Ba^{++} are present in the peripheric head group region, water molecules hydrogen bonded to hydrophilic groups are displaced with the result of a tighter packing throughout the bilayer and hence a decrease in permeability of the cell membrane.

Thus, there is an interdependence between the packing density of the hydrocarbon tails and that of the hydrophilic head groups. The interactions of water with the latter loosens the packing and this manifests itself in an increase in motion of the phosphoryl choline groups. Simultaneously, the motional freedom of the hydrophobic chains and the permeability of the membrane is increased with simultaneous decrease in T_c.

Phospholipid bilayer + water : looser packing and increase in permeability

Phospholipid bilayer + cations: tighter packing and decrease in permeability

An extrusion of water from the hydrophilic group region is also accomplished by the incorporation of cholesterol.

The innermost hydration shell contains one tightly bound water molecule with a rotational correlation time of 10^{-7} seconds. The main hydration shell consists of 11 to 12 water molecules with a rotational correlation time of about 8.10^{-10} seconds. In addition, there are for each lipid molecule about 11 to 13 molecules "trapped" water with a correlation time of 3.10^{-10} seconds [1].

The amount of bound water (average value 12.5 mol water per mol phosphatid choline) is consistent with the quantity of non-freezable water derived from calorimetric measurements.

The penetration of water through the membrane is described in two different ways:

1. by bulk flow,

2. by molecular flow (diffusional, molecule by molecule).

Across artificial non-liquid membranes (cellophane) water is considered to move almost exclusively by bulk flow.

Transport of a solute across the membrane is called active, if it occurs *"uphill"*, i.e. against the concentration gradient and if the solute is charged (ionic) against an electric field, i.e. from a lower to a higher thermodynamic potential (*violating the simple chemical and physical laws*).

Little is known about the mechanism of electrolyte transport, apart from the likely participation of the transport adenosine triphosphatase. An older, but revived hyphotheses is that of a redox-pump.

5. The Role of Membranes in the Organism

Water penetrates not only the cell membranes but also all other parts of the organism to different degrees and in this way *water is in continuous interactions with all parts of the organism and hence it provides for the unity of the organism on its material level.*

The *"one-ness" of the human organism* with its 100 trillions of cells (each of them an individuality, but all of them in continuous cooperative interactions) originated from one cell with its water content. The human ovum is a unicellular being consisting of a nucleus, surrounded by the cell membrane and in between them the connecting aqueous cytoplasm.

The highly differentiated and highly developed cell membrane is fluid and deformable, it withstands actions from both sides and at the same time coordinates all of the activities with the help of the hydrophilic regions into the different and separated extracellular and intracellular regions respectively (see p. 209 ff).

The intracellular region retains the individuality of the cell, whereas the extracellular regions are individually less differentiated. The former act "structurizing" for the differentiation within the organism and the latter give unity to the organism, whereby each cell serves an individual purpose within the whole in co-operation with all of the other cells.

This requires that

(i) certain characteristic features are maintained and these are usually considered as *conservative structures* and

(ii) certain changes and alterations are taking place with the development of new structural features and these are manifestations of developing structures referred to as *dissipative structures* or *non-equilibrium structures*.

The former are considered to be formed and maintained through reversible transformations implying no appreciable deviation from equilibrium. The latter are formed and maintained through the effect of exchange of energy and matter under non-.equilibrium conditions [9].

The formation of cell patterns at the onset of free convection is a typical example of a dissipative structure [10]. A convection may be considered as a *"giant fluctuation* stabilized by the flow of energy and matter prescribed by the *conservative boundary conditions"*[11].

Lewis [12] proposed to unify fluctuation theory and equilibrium thermodynamics. However, he was concerned only with equilibrium situations, where the effect of fluctuations is generally negligible (with exception of critical phenomena).

A distinction is made between an equilibrium state and a *steady state*, the former being characterized by a "zero-entropy production" and the latter with non-vanishing entropy production [11].

In metastable states, the system is considered to be prevented from reaching equilibrium. It is therefore not possible to draw a clear border line between conservative boundary conditions and developing dissipative structures, as both of them are necessary. In the complete absence of dissipative features, a conservative structure would be unobservable as each dissipative structure requires conservative aspects in order to be perceptible.

A conservative structure provides certain boundary conditions for the development of the dissipative features and these help at the same time to provide for the said conservative boundary conditions to which they yield. The necessity of conservative boundary conditions for dissipative processes and their mutual interactions may be illustrated by considering a river.

The conservative boundary conditions for a river are provided by embankment, river bed and gravitational forces. River bed and embankment are slightly, but continuously altered, even if the speed and the quantity of water remain nearly constant. It is altered increasingly, the greater the changes in the rate of flow. With a reduction of the quantity of flowing water, the river bed will be silted up, while an increase will cause the river to overflow, the bed to deepen and the banks to change their shape. In other words: Any change in dissipative features (rate of flow) causes changes in conservative features (river bed), and these in turn have an influence on the dissipative aspects. Thus, there are always mutual interdependencies between dissipative and conservative features. The former follow the latter, and these provide boundary conditions for the former.

It is characteristic for dissipative features that they have a direction of flow that is influenced by conservative boundary conditions. River-regulations, elimination of cascades, damming up for power stations etc. are known to have a decisive influence on the dissipative features of the river. Even within the dissipative features conservative elements are present, for example, stone and sand transported by the river or the communication of flowing water with ground water.

This shows again that

(i) dissipative features require conservative boundary conditions without being conservative themselves, and

(ii) *knowledge about static aspects would be impossible in the complete absence of dynamic aspects.*

It has been said that the cell membranes are part of the water system of the organism (see p. 209 ff) and this means that they are produced by the system as conservative boundary conditions, which may be changed according to the requirements. At the same time, the conservative boundary conditions must be fully respected by the dissipative actions of the system: *dissipative features appear to determine the conservative aspects and the conservative aspects appear to condition the dissipative aspects.*

The consideration of dissipative processes in their qualitative and quantitative aspects show that they serve a certain purpose in the organism. In the absence of dissipative processes, the energy balance could not be maintained within the organism. For this purpose the production of ATP is required, as well as its availability in certain areas, in certain quantities and at certain times, according to the necessities for the maintenance of the whole organism.

We wish to point out that the widely used term "self-organization" implies the idea of an accidental origin. The use of the prefix "self" presupposes a "self" as a "being" , which in itself represents the completeness of the full program. Organization is always serving a certain purpose to which the parts yield, and hence it cannot be produced by the actions of the parts themselves, much less by chance. It is the "self" within the organization that makes the parts become organized in certain ways. The "self" is therefore not a consequence but rather a prerequisite for the organization.

The enormous influence of the conservative boundary conditions on the development of dissipative aspects shows that *a dissipative structure cannot develop on its own*. It depends on the available potentialities, which are developing only in the presence of conservative boundary conditions. These represent aspects of a certain "building plan" that cannot be designed by the developing system itself. We suggest therefore to replace the term "self-organization" by referring to a *"self-performance" according to a plan*.

References

1. H. Hauser, in *Water a Comprehensive Treatise*, ed. F. Franks, Vol. 4, Chapter 4, p. 209, (Plenum Press, New York, 1975).
2. G. C. Kresheck, in *Water a Comprehensive Treatise*, ed. F. Franks, Vol. 4, Chapter 2, p. 95, (Plenum Press, New York, 1975).
3. J. M. Corkill and J. F. Goodman, *Adv. Coll. Interface Sci.* **2** (1969) 297.
4. E. J. Fendler and J. H. Fendler, *Adv. Phys. Org. Chem.* **8** (1970) 271.
5. M. J. Schick (ed.), *Nonionic Surfactants*, (Marcel Dekker, New York, 1971).
6. D. W. Wetlaufer, S. K. Malik, L. Stoller and R. L. Coffin, *J. Amer. Chem. Soc.* **86** (1964) 508.
7. K. Shirahama, M. Mayashi and R. Matuura, *Bull. Chem. Soc. Japan* **42** (1969) 1206, 2123.
8. C. Tanford, *The Hydrophobic Effect: Formation of Micelles and Biological Membranes* (Krieger Publ. Co. Malabar/Florida, USA 1991).
9. G. N. Lewis and M. Randall, *Thermodynamics*, Revised by K.S. Pitzer and L. Brewer, 2nd ed. (McGraw Hill Book Co., New York 1971).
10. H. Benard, *Rev. Gen. Sci, Pure Appl.* **11** (1900) 1261.
11. P. Glanssdorff and I. Prigogine, *Thermodynamic Theory of Structure, Stability and Fluctuations*, (Wiley-Intersci., London, New York, 1971).
12. G. N. Lewis, *J. Amer. Chem. Soc.* **53** (1931) 2578.

CHAPTER 10

HYDROPHOBIC SOLUTES IN WATER

1. Solubility Considerations

It has been described in the preceding chapter that large molecules with a hydrophilic group and one or two long hydrophobic chains are hardly soluble in water and tend either to occupy positions at the interface area or to form micelles and bilayers.

Likewise, the solubilities of hydrophobic solutes, the molecules of which have no hydrophilic group, are very small. For example, the solubility of hydrophobic pentane is by three orders of magnitude smaller than that of 1-pentanol, which has a hydrophilic group.

It is interesting to note that the solubility of water in the hydrophobic liquid is usually by at least one order of magnitude smaller than the solubility of the hydrophobic liquid in water (Table 14).

Table 14. Solubilities of hydrophobic liquids in water and of water in hydrophobic liquids.

Hydrophobic Liquid	Mole Fraction of Hydrophobic Liquid in Water	Mol Fraction of Water in the Hydrophobic Liquid
pentane	$9.5 \cdot 10^{-6}$	$4.8 \cdot 10^{-4}$
hexane	$2.0 \cdot 10^{-6}$	$3.5 \cdot 10^{-4}$
heptane	$6.0 \cdot 10^{-7}$	$4.6 \cdot 10^{-4}$
benzene	$4.1 \cdot 10^{-4}$	$2.7 \cdot 10^{-3}$
toluene	$1.0 \cdot 10^{-4}$	$1.7 \cdot 10^{-3}$
1,2-dichloroethane	$9.2 \cdot 10^{-4}$	$1.1 \cdot 10^{-2}$
chloroform	$1.2 \cdot 10^{-3}$	$4.8 \cdot 10^{-2}$
carbon tetrachloride	$9.0 \cdot 10^{-5}$	$5.4 \cdot 10^{-4}$

The solubilities of gases are about in the same order of magnitude and listed in Table 15. The higher solubilities of sulphur dioxide and of carbon dioxide are due to their interaction with water to form ionic species.

The solubilities of gases in water are smaller than in most other solvents (Table 16). On the other hand, it is not possible to remove the last traces of dissolved gases from water or from any other solvent. Carefully degassed water still contains about 10^{-6} mol gas per litre and this concentration is by one order of magnitude greater than that of hydrated protons in water produced by self-ionization (Chapter 7).

Table 15. Gas solubilities in water at 25°C.

Gas	Mole Fraction	Gas	Mole Fraction
SO_2	$2.46 \cdot 10^{-2}$	O_2	$2.29 \cdot 10^{-5}$
CO_2	$6.15 \cdot 10^{-4}$	CO	$1.77 \cdot 10^{-5}$
Xe	$7.89 \cdot 10^{-5}$	H_2	$1.41 \cdot 10^{-5}$
Kr	$4.51 \cdot 10^{-5}$	N_2	$1.18 \cdot 10^{-5}$
CH_4	$2.55 \cdot 10^{-5}$	Ne	$8.15 \cdot 10^{-6}$
Ar	$2.52 \cdot 10^{-5}$	He	$7.00 \cdot 10^{-6}$

Table 16. Solubilities of hydrophobic gases in different solvents in molar fractions at 25°C.

Solvent	N_2	O_2	Ar	CH_4	H_2
Water	$1.20 \cdot 10^{-5}$	$2.30 \cdot 10^{-5}$	$2.50 \cdot 10^{-5}$	$2.60 \cdot 10^{-5}$	$1.40 \cdot 10^{-5}$
Methanol	$2.36 \cdot 10^{-4}$	$4.12 \cdot 10^{-4}$	$4.45 \cdot 10^{-4}$	$9.19 \cdot 10^{-4}$	$1.57 \cdot 10^{-4}$
Ethanol	$3.44 \cdot 10^{-4}$	$5.79 \cdot 10^{-4}$	$6.20 \cdot 10^{-4}$	$1.30 \cdot 10^{-3}$	$2.14 \cdot 10^{-4}$
1-Propanol	$4.06 \cdot 10^{-4}$	$5.03 \cdot 10^{-4}$	$7.76 \cdot 10^{-4}$	$1.56 \cdot 10^{-3}$	$2.28 \cdot 10^{-4}$
1-Butanol	$4.61 \cdot 10^{-4}$	$7.89 \cdot 10^{-4}$	$9.09 \cdot 10^{-4}$	$1.91 \cdot 10^{-3}$	$2.67 \cdot 10^{-4}$
DMSO	$0.83 \cdot 10^{-4}$	$1.96 \cdot 10^{-4}$	$1.54 \cdot 10^{-4}$	$3.86 \cdot 10^{-4}$	$7.60 \cdot 10^{-5}$
Aceton	$5.42 \cdot 10^{-4}$	$6.60 \cdot 10^{-4}$	$9.06 \cdot 10^{-4}$	$1.88 \cdot 10^{-3}$	$3.01 \cdot 10^{-4}$
Benzene	$4.24 \cdot 10^{-4}$	$7.58 \cdot 10^{-4}$	$8.85 \cdot 10^{-4}$	$1.77 \cdot 10^{-3}$	$2.59 \cdot 10^{-4}$
Nitrobenzene	$2.64 \cdot 10^{-4}$	$4.95 \cdot 10^{-4}$	$4.45 \cdot 10^{-4}$	$1.06 \cdot 10^{-3}$	$1.56 \cdot 10^{-4}$
CCl_4	$6.40 \cdot 10^{-4}$	$9.74 \cdot 10^{-4}$	$1.36 \cdot 10^{-3}$	$2.92 \cdot 10^{-4}$	$3.19 \cdot 10^{-4}$

Aqueous solutions of gases reveal peculiarities not observed in other liquids [1,2]. For example, most gaseous solutes show a partial molar entropy and enthalpy of dissolution substantially lower than other solutes and a partial molar heat capacity, which is anomalously large [1]. One of the difficulties for the thermodynamic description of aqueous solutions of gases are their low solubilities, which render direct calorimetric studies very difficult. It is for this reason that thermodynamic quantities have been extracted almost exclusively from solubility data.

The solubilities of hydrophobic gases and of hydrophobic alkanes show negative temperature coefficients. Alkyl and aromatic compounds with only one or at least two hydrophilic groups show also hydrophobic properties. This group of solutes includes also alkyl- or aryl substituted complex ions, such as tetrabutyl ammonium or tetraphenyl borate ions as well as a wide range of different polymers.

2. Structural Aspects

One of the first rigorous approaches to the problem of gas solubility in water was made by *Eley* [3,4]. He promoted the idea of considering the process of dissolution as a consequence of two consecutive processes:
(i) Cavity formation and
(ii) Solvent - solute interactions.

This approach led to large molar heat capacities of the solutes and to the conclusion that these promote the water structure [5-7]. This necessitates that the solute is well-integrated in the solution structure and modified itself by the liquid [8].

Water was found to form crystalline hydrates with xenon, chlorine, methane and other gases and these are known as clathrate hydrates. The results of X-ray diffraction experiments revealed that - with a few exceptions - the structural feature common to them is the pentagonal dodecahedron of water molecules [9-12].

Two types of gas hydrates are distinguished [9]:
(i) Structure I contains 46 water molecules in the unit cell with a lattice constant of 1200 pm. It contains six big holes up to 590 pm diameter and two small holes. The latter are located at the vertices and the centre of the unit cell. Each of them are formed by a pentagonal dodecahedral array of 20 water molecules. These small cavities are empty. The remaining water molecules in the unit cell form bridges between these dodecahedra in such a way that a second type of cavity is provided [11].

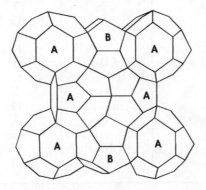

Fig. 23. Idealized gas hydrate structure type I, A: tetrakisdecahedron, B: pentagonal dodecahedron [11].

(ii) Structure II has a unit cell with a lattice constant of 1730 pm. It contains 136 water molecules arranged around 20 great holes of 590 to 690 pm diameter and 8 smaller holes. The small holes are usually empty, but sometimes they may contain small molecules such as nitrogen, oxygen or carbon dioxide. If they are empty and the big holes occupied, the ideal composition is 17 water molecules for each gas molecule.

If air is present in the tetrahydrofuran structure II clathrate hydrate at atmospheric pressure, the solute molecule is entirely taken into the small cages of the hydrate with slight increase in density and slight increase in melting point from 4.40°C to 4.45°C. On gentle warming the frozen sample, bubbles are evolved [13].

Because of the stabilizing effect of a second gaseous encageable component, the latter have been termed "help gases" [14,15]. Their so called "structure promotion effect" has also been established for aqueous solutions [14] and this suggests that dissolved gases are capable of stabilizing water structures [16].

Table 17. Stabilizing effects by help gases [15].

Clathrate Former	Decomp. temp.[°C]	Help Gas	Pressure [atm]	Decomp. temp. [°C]
C_2H_5Cl	4.8	CO_2	5	6.8
			15	10.8
		N_2	5	6.0
			15	8.6
$CHCl_3$	1.7	CO_2	5	5.9
			15	10.5
		N_2	5	3.6
			15	6.5
CH_2Cl_2	1.7	CO_2	5	5.0
			15	8.4
		N_2	5	3.1
			15	6.2
C_2H_5Br	1.4	CO_2	5	4.9
			15	9.1
		N_2	5	2.8
			15	5.6
CCl_4	No hydrate*	CO_2	5	7.3
			15	12.6
		N_2	5	3.1
			15	7.0

* Hydrate stable only in the presence of "help gas"

Detailed structural studies revealed that not only hydrophobic molecules, but also hydrophobic *ions* may be situated in cages. Examples are provided by *n*-butylammonium benzoate [17], tetra *n*-butylammonium fluoride [18] and tetra *iso*-amylammonium fluoride [19]. In the latter clathrate part of the ideal hexagonal structure is distorted and the N and F atoms replace water molecules of the undistorted structure (Fig. 24).

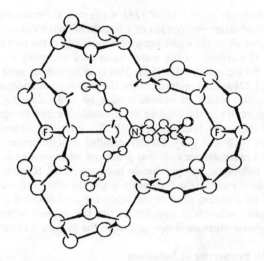

Fig. 24. Structure of tetra *iso*-amylammonium fluoride clathrate.

The cages are longer than in the idealised structure in order to meet the spatial requirements of the ions. Because of the hydrophilic properties of the nitrogen atom and in particular of the fluoride ion, these are anchored within the hole boundaries by replacing water molecules. For these reasons, the tetraalkylammonium ions have no longer rotational freedom as is characteristic for dissolved gas molecules.

Considerations about the cavity energy have been made by *Emi* and *Bockris* [20]. They suggested to estimate the cavity energy from the cavity area and the specific surface free energy. As values for both surface tension and gas solubilities show negative temperature coefficients, cavity stabilities are increased as the temperature is lowered. Their stabilities are greater the lower the temperature and the greater the surface area. Due to the surface tension a certain pressure, the so-called capillary pressure, is exerted from the interface towards the internal regions. This requires a certain "counter-pressure" from inside and this appears to be provided by the cages. This is why, decrease in drop size of the liquid is paralleled by an increase in hole concentration.

Buckingham [21] estimated a value of 10 kcal/mol for the energy of cavity formation. *Mayer* [22] has shown that the energy of cavity formation is related to the value of the molar enthalpy of evaporation of liquid water at 25°C and hence to the surface tension at this temperature. He also found relations between surface tension and enthalpy of evaporation for a variety of solvents [22].

Mention has been made that to a certain extent the size and the shape of the holes appears to be adaptable to the requirements of the solute particles. *King* [23] has shown that there is no inherent direct limitation for the size of the polyhydra in the structural networks. *Hagler et al* [24] suggested a continuous distribution of cluster sizes in liquid water on grounds of a statistical thermodynamic treatment.

In the clathrates all of the hydrogen atoms available in the unit cell are hydrogen bonded with O - O distances of the water framework enclosing a cavity shorter for unoccupied than for occupied holes [25]. This is in agreement with the bond length variation rules [26,27] and a requirement for the existence of unoccupied holes.

According to the results of optical studies on molecular oxygen dissolved in water, oxygen molecules in the cavities act as weak electron acceptors towards the inner surfaces of the water structure [28,29]. The following interesting observation has been reported by *Evans* [30]: Freshly distilled dimethylamine quickly becomes yellow in air and a much deeper colour is produced when oxygen is bubbled through the liquid. Such coloration has previously been attributed to permanent oxidation. However, the colour completely disappears when the dissolved oxygen is removed by evacuation or by a stream of nitrogen. Also, the new absorption band developed by various aromatic substances (usually in the UV region) when saturated with oxygen at atmospheric pressure disappears when the oxygen is removed.

3. Structures and Properties of Solutions

The dissolved gas molecules show an exceptionally high degree of rotational mobility [31]. Interesting conclusions have been reached from ^{13}C NMR spin-lattice relaxation time measurements on aqueous solutions of propanol and *tert*-butanol. According to these water is rather unusual

(i) in the extent to which it restricts the motion of hydrophobic solute moieties and

(ii) in conjunction with the concomitant reduction in the motion of neighbouring water molecules [32-34].

The remarkable mutual adaptabilities of water structure and dissolved gas molecules implies that information between them is continuously exchanged. Results of studies on heavy water have been taken as suggesting that "structure promotion" by non-polar solutes is greater in D_2O than in H_2O [25]. Pure liquid D_2O is believed to possess a higher degree of hydrogen bonding in H_2O, and hence to be more structured [1,35,36].

The structural features around a hole of the dimensions as found in the clathrate structures have been illustrated by applying the bond length variation rules [27,37] and they are presented in Fig. 25 [38,39].

After addition of a hydrophobic solute the vibration spectrum of the solution is changed in the same way as it results from an increase in temperature of "pure" water. This means that in the presence of hydrophobic solutes the water structure is "loosened", i.e. becoming more "water-like" and hence these solutes are also called "structure makers".

The "loosening" and the "dynamization" of the water structure is reflected in the changes in macroscopic properties which are different from those caused by the dissolution of hydrophilic solutes.

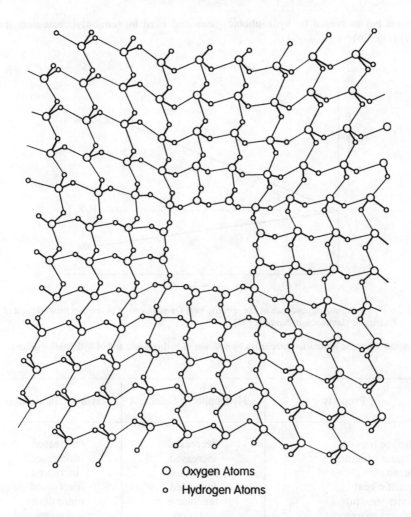

Fig. 25. Idealized two-dimensional illustration of the water structure around a hole [38,39].

It is remarkable that addition of hydrophilic solutes leads to changes in solution properties which are opposite to those produced by addition of hydrophobic solutes.

Whereas the vapour pressure is decreased by hydrophilic solutes, it is increased by the presence of dissolved gases (Table 18). The surface tension is increased by hydrophilic solutes, but decreased by hydrophobic solutes. Remarkable are also the different effects on the values for the heat capacity. This is decreased by hydrophilic

solutes but increased by hydrophobic gases and even by tetrabutylammonium ions
[40] (Fig. 26).

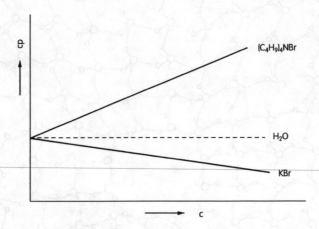

Fig. 26. Concentration dependence of the specific heat for solutions of potassium bromide and of
tetrabutylammonium bromide [40].

Table 18. Changes in solution properties by addition of hydrophobic and of hydrophilic solutes
to water.

Property	Hydrophobic Solutes	Hydrophilic Solutes
Surface tension	decreased	increased
Vapour pressure	increased	decreased
Density	decreased	increased
Specific heat	increased	decreased
Water structure	less dense	more dense
Particle movement	rotation	one-dimensional
Vibration spectra correspond to those of pure water at	higher temperature	lower temperature
Actions on the static aspects of "water structure"	decreasing	increasing
Actions on the dynamic aspects of "water structure"	increasing	decreasing

The structural aspects in solutions of clathrates appear also similar to those in the crystalline compounds. A study of the alkyl proton resonance's of symmetrical tetraalkylammonium ions [41] showed that the infinite dilution values, which are much higher in water than in acetonitrile, appear to be the result of the collective properties of the water structure with a relatively stable configuration of the water molecules around the non-polar alkyl chains.

This is also evidenced by the temperature dependence of the proton chemical shift, which is decreased by increase in temperature: increased thermal motion causes progressive destruction of the enhanced water structure around the hydrophobic chains. The different protons of the alkyl groups are not equally effected by changes in temperature and in concentration with greater shifts of the inner CH_2-group protons. This suggests that the positive charge of the nitrogen atom is distributed over the neighbouring methylene groups, which is in accordance with quantum mechanical requirements [42].

At low concentrations, the slight changes in δ show a weak increase for Et_4NBr, nearly constant values for Pr_4NBr and even a slight decrease for Pr_4NF and Bu_4NBr. In this range cation-cation interactions are dominant. Whereas in solutions so concentrated that there is not enough water available for the required structural arrangement which provides holes for the pairs of tetraalkylammonium ions, cation-anion interactions become inevitable. $\Delta\delta$ values for the fluoride remain well below those of the bromide, which in view of the greater effective polarizating power of a gaseous fluoride ion as compared to that of the bromide ion, is interpretable by a considerably stronger solvation of the fluoride ion incorporated within the aqueous framework similar to that in the corresponding clathrate hydrate. Hence it is much easier for the bromide ion to interact with the cation.

An important finding is the effect of the amount of dissolved air on the surface tension of aqueous solutions [43]. A change with time was found to depend strongly on the initial quantity of air dissolved in the water sample. Surface tension is related to the structural differences of the interfacial region relative to the bulk.

4. Dynamic Features

Whereas ions are known to move in one dimension, the dissolved gas molecules have - although within the holes - a certain freedom in all directions. This is because the size of the holes for the gas molecules is much larger than required for their incorporation. This means that the gas molecules can move within the holes [31] and this is in agreement with the high values for the heat capacities of the gas molecules, which are nearly as high as in the gas phase.

This shows that the gas molecules within the cages are nearly as mobile as in the gaseous state, but within the holes their motions are somewhat limited by their continuous interactions with the flexible inner surfaces, which provide a kind of static boundary conditions for their motions.

Accordingly, the motions of the dissolved gas molecules are described as rotations or as librations [31,44]. Their amplitudes are confined by the flexible and yet resistant inner surfaces of the holes in the liquid. This means that the oscillations of the gas molecules are influenced by the motion pattern of the inner surfaces, which represent characteristic features of the oscillating pattern of the whole liquid. In this way the oscillations of the gas molecules are modified until they come in harmony with the oscillating pattern of the liquid, which is modified itself.

References

1.   E. Wilhelm, R. Battino and R. J. Wilcock, *Chem. Rev.* **72** (1972) 211.
2.   E. Wilhelm and R. Battino, *Chem. Rev.* **73** (1973) 1.
3.   D. D. Eley, *Trans. Farad. Soc.* **35** (1939) 1281, 1421.
4.   D. D. Eley, *Trans. Farad. Soc.* **40** (1944) 184.
5.   J. L. Kavanau, *Water and Solute-Water Interactions,* (Holden-Day, San Francisco 1964).
6.   A. Yu. Namiot, *J. Struct. Chem.* **2** (1961) 381, 444.
7.   P. Krindel and I. Elizier, *Coord. Chem. Rev.* **6** (1971) 217.
8.   V. Gutmann and G. Resch, *Pure Appl. Chem.* **53** (1981) 1547.
9.   M. v. Stackelberg and H. R. Müller, *Z. Elektrochem.* **58** (1954) 25.
10. M. v. Stackelberg, *Naturwiss.* **36** (1949) 327, 359.
11. G. A. Jeffrey and R. K. Mc Mullan, *Progr. Inorg. Chem.* **8** (1967) 43.
12. V. Gutmann, E. Plattner and G. Resch, *Chimia* **31** (1977) 431.
13. M. v. Stackelberg and J. Frühbuss, *Z. Elektrochem.* **58** (1954) 99.
14. D. W. Davidson, in *Water, a Comprhensive Treatise,* ed. F. Franks, Vol. 2, Chapter 3, p. 115 (Plenum Press, New York 1973).
15. M. v. Stackelberg and W. Meinhold, *Z. Elektrochem.* **58** (1954) 40.
16. W. Drost-Hansen, *Fed. Proc.* **30** (1971) 1539.
17. M. Bonamico, G. A. Jeffrey and R. K. Mc Mullan, *J. Chem. Phys.* **37** (1962) 2219.
18. R. K. Mc Mullan, M. Bonamico and G. A. Jeffrey, *J. Chem. Phys.* **39** (1963) 3295.
19. D. Feil and G. A. Jeffrey, *J. Chem. Phys.* **35** (1961) 1863.
20. T. Emi and J. O'M. Bockris, *J. Phys. Chem.* **74** (1970) 159.
21. A. D. Buckingham, *Disc. Farad. Soc.* **24** (1957) 151.
22. U. Mayer, *Monatsh. Chem.* **109** (1978) 421.
23. R. B. King, *Theor. Chim. Acta* **25** (1972) 309.
24. A. T. Hagler, H. A. Scheraga and G. Nemethy, *J. Phys. Chem.* **76** (1972) 3229.
25. D. W. Davidson, *Can. J. Chem.* **49** (1971) 1224.
26. V. Gutmann, *Electochim. Acta* **21** (1976) 661.
27. V. Gutmann, *The Donor - Acceptor Approach to Molecular Interactions,* (Plenum Press, New York, 1977).
28. L. J. Heidt and A. M. Johnson, *J. Amer. Chem. Soc.* **79** (1957) 5587.
29. J. Jortner and U. Sokolov, *J. Phys. Chem.* **65** (1961) 1633.
30. F. D. Evans, *J. Chem. Soc.* (1953) 345.
31. J. W. Tester, R. L. Virins and C. C. Herrick, *AIChE J.* **18** (1972) 1220.
32. O. W. Howarth, *J. Chem. Soc. Faraday Trans.* **1** (1975) 71.
33. T. S. Sharma and J. C. Abluwalia, *Chem. Soc. Rev.* **2** (1973) 203.

34. E. V. Goldhammer and H. G. Hertz, *J. Phys. Chem.* **74** (1970) 3734.
35. G. C. Kresheck, H. Schneider and H. A. Scheraga, *J. Phys. Chem.* **69** (1965) 3132.
36. G. Nemethy and H. A. Scheraga, *J. Phys. Chem.* **41** (1964) 680.
37. G. Resch and V. Gutmann, in *Advances in Solution Chemistry,* eds. I. Bertini et al, p. 1 (Plenum Press, New York 1981).
38. V. Gutmann, E. Scheiber and G. Resch, *Monatsh. Chem.* **120** (1989) 671.
39. V. Gutmann, G. Resch and E. Scheiber, *Revs. Inorg. Chem.* **11** (1991) 295.
40. K. Trincher, *Water Research* **15** (1981) 433.
41. G. Kabisch, *Ber. Bunsenges. Phys. Chem.* **80** (1976) 602.
42. G. N. J. Port and A. Pullmann, *Theoret. Chim. Acta* **31** (1973) 231.
43. H. Sobol, G. Garfias and J. Keller, *Z. physik. Chem.* (n.F.) **80** (1976) 1941.
44. J. H. van der Waals and J. C. Platteeuw, in *Advances in Chemical Physics,* ed. I. Prigogine (Intersci. Publ. New York 1959).

CHAPTER 11

HYDROPHILIC SOLUTES IN WATER

1. Hydrated Ions in Pure Water

An enormous number of results is available on the structures, thermodynamics, kinetics and theory of solutions of hydrophilic solutes in water (with emphasis on electrolyte solutions). All of the dissolved ions or molecules are hydrated by water molecules and hydration structures have been studied in considerable detail.

The self-ionization equilibrium of liquid water has been investigated and its significance is realized for pH-considerations and the actions of buffer solutions. The hydration structures of ions produced by self-ionization have been carefully investigated, but their significance for the existence and properties of "pure" water has not been mentioned until recently [1].

Apart from the unavoidable presence of the hydrated ions produced by self-ionization, it proved impossible to remove the last traces of other solutes as a consequence of the continuum of matter, as outlined in some detail in Chapter 2. For example, it proved impossible to remove the last traces of hydrogen carbonate ions in the course of purification for the production of the so-called "Kohlrausch-water".

All of the unremoveable solutes contribute to the structural and energetic differentiation within the liquid, which could never exist as a uniform dielectric medium, and hence they can be taken as being required for the existence of the liquid.

Most of the efforts in studies of solutions are directed to the modifications of the dissolved solutes, but their functions as centres for the modification of the liquid structure are investigated only as far as the changes within the "hydration shells" can be detected.

Hydrophilic solutes are also called *structure breakers*, because they interact with water molecules more strongly than do water molecules with each other. IR- and NMR-spectra of solutions of non-electrolytes, such as HMPA, indicate changes which simulate the effects of a decrease in temperature of the pure water spectrum [2]. What are called "thermal vibrations" are actually complicated dynamic patterns with a character all of their own that reflects the energetic influence by the environment, including that of temperature, as well as those of pressure, irradiation, fields, mechanical forces, drop size etc.

2. Structure of Hydrated Hydrogen Ions

Bond lengths and bond angles around an atomic hydrogen ion (which in fact cannot be recognized as such because it exists only within the complex liquid structure) are highly influenced by the number of water molecules attached and by the nature and number of other solute particles. For example, in crystalline

$HClO_4 \cdot H_2O$ the three hydrogen atoms originating from the central oxygen atom are coordinated to perchlorate groups, but the O - O distances are slightly different, varying in mean values between 263 and 271pm [3] (Fig. 27).

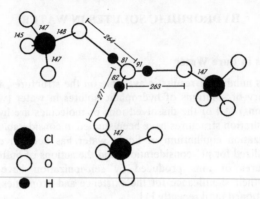

Fig. 27. Crystal structure of $HClO_4 \cdot H_2O$.

The structure of $HClO_4 \cdot 2\,H_2O$ may be derived from that of the monohydrate by replacement of one of the weakly coordinated perchlorate groups by a strongly donating water molecule which is hydrogen bonded by one of the three hydrogen atoms of the H_3O-group to give the H_5O_2-group. Thus, the O - O distance between the two water units is shorter, namely 242 pm and the hydrogen bond more symmetrical than in the monohydrate. On the other hand the O - O distances connecting oxygen atoms of the H_5O_2 unit to perchlorate oxygen atoms are somewhat longer, 278 and 279 pm respectively [4] (Fig. 28).

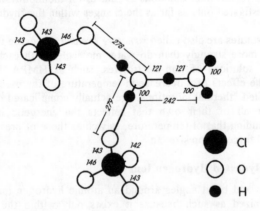

Fig. 28. Crystal structure of $HClO_4 \cdot 2\,H_2O$.

The classical $H_9O_4^+$ ion is represented by a central H_3O^+ component pyramidically surrounded by three water molecules with an average O - O bond distance of 257 pm and an average O - O - O bond angle of 111° [4]. The hydrogen bond distances and angles deviate due to differences in crystallographic environment of the water molecules surrounding the H_3O^+ unit.

The largest protonated unit that has been structurally characterized is the $H_{13}O_6^+$ unit, which has a high symmetry when symmetrically surrounded by chloride ions [5].

According to the conventional approach the chemical properties of aqueous ions are related to the properties of the analogous non-hydrated (hypothetical) ions [6,7], although it is well-known that it is not possible to assign thermodynamic properties to individual ions without making extra-thermodynamic assumptions. Such assignments cannot be checked directly and they vary with the theoretical approach. Furthermore, idealized ions cannot be found in the liquid, in which they are "embedded" without observable border lines.

It is well-known that the thermodynamic activity of an ionic species depends not only on the presence of other ions, but also on the concentration of the ionic species under consideration. This means that there is also a concentration effect on the structure within the hydration shells. Ion-ion interactions have been studied by means of various spectroscopic techniques [8], but interactions between hydrophilic and hydrophobic solutes have hardly been given the attention that they deserve.

3. The Donor-Acceptor Approach to the Hydration of Ions

A generally valid qualitative description is provided by the extended donor-acceptor approach. By the hypothetical addition of water molecules to the gaseous ion to give the first hydration layer, the positive net charge of the cation is decreased and the O-H bonds of the water molecules are lengthened with slight increases in the positive net charges at the peripheric hydrogen atoms.

The decrease in net charge at M is equal to the sum of the increases in positive net charges of all of the hydrogen atoms of the water molecules in the first hydration sphere. In other words: the loss in positive net charge at M is spread over the peripheric hydrogen atoms [9]. The increase in acceptor properties of the hydrogen atoms favours the formation of the second hydration sphere. In this way the net charge at M is further decreased and the M-O bond is shortened. The O-H bonds of the water molecules in the first sphere are lengthened just as well as in the second sphere. The total loss of positive charge of M is redistributed over the peripheric hydrogen atoms of the water molecules in the second hydration sphere [9,10].

The greater the number of hydration layers, the shorter is the M-O bond and the smaller the net positive charge residing at the cation as a result of its modification by the hydrated water molecules [8].

$\longrightarrow$ increasing O·····O distances

The shorter the distance of the water molecules from the ion, the shorter are the O - O distances and hence the greater the local density. The greater the distance from the ion, the greater is both the asymmetry and the deviation from linearity of the hydrogen bonds between the water molecules. Each water molecule has therefore a different environment and hence different properties. The static aspects may be considered at any time by a characteristic pattern of inhomogeneties that is established throughout the liquid system and influenced locally by the so-called solute-solute interactions, mediated through the solvent molecules (Fig. 29).

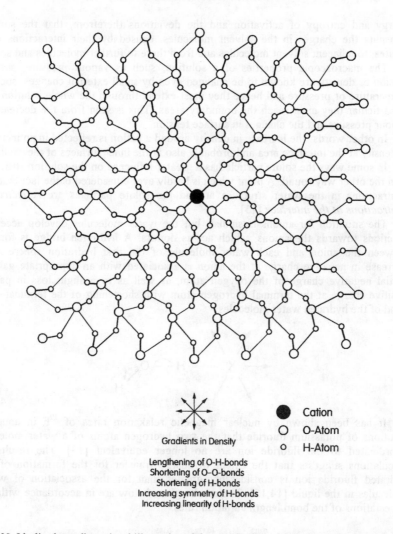

Gradients in Density

● Cation
○ O-Atom
○ H-Atom

Lengthening of O-H-bonds
Shortening of O-O-bonds
Shortening of H-bonds
Increasing symmetry of H-bonds
Increasing linearity of H-bonds

Fig. 29. Idealized two-dimensional illustration of the modification of the water structure around a cation.

Each ion is therefore modified by and modifying the solvent structure. The ions are therefore analogous to interstitial positions in solids [11,12]. All of the solutes are in continuous interactions with the whole water structure.

The long range effects by an ion on the water structure are reflected in various facts. *Likhtenstein* [13] concluded from the linear relationships between activation energy and entropy of activation and the deviations therefrom, that the solvent transmits the charges in the solvent molecules, caused by their interactions with solutes, to adjacent solvent molecules and from there to further molecules and so on.

The macroscopic properties of a solution, such as vapour pressure, surface tension or density are known to be changed even by small external changes, such as temperature or pressure and hence they must extend through the whole solution and in particular they must reach the boundary areas. This is seen from the decrease in vapour pressure and the changes in surface tension.

In other words: the increase in density around a cation is reflected in an increase in density at the interface area and probably also at the inner surfaces of the cavities.

In some ways the solute particle has a greater influence on the solution structure than the other way round. A point which is hardly ever considered is the fact that the contractions in the water structure around the solute particles are *reflected in contractions at the interface* [15].

The situation for anions is similar, but the water molecules develop acceptor functions towards the anions, which act as donors. A hydrogen bridge is formed between the anion and each water molecule of the first hydration sphere. The decrease in negative charge at the anion is associated with an appropriate gain in partial negative charge of the oxygen atom, as well as to a slight loss in partial positive charge at the terminal hydrogen atom with shortening of the terminal O-H bond of the hydrated water molecules.

It has been shown by nuclear magnetic relaxation rates of ^{19}F in aqueous solutions of potassium fluoride that the two hydrogen atoms of a water molecule coordinated to the fluoride ion are no longer equivalent [13]. The results of calculations suggests that the extent of charge transfer for the formation of the hydrated fluoride ion is considerably greater than for the association of water molecules in the liquid [14,15]. The results given below are in accordance with the expectations of the bond length variation rules.

As the donor properties of the oxygen atom have been increased in this way, a second hydration sphere is readily formed, and the X-H bond is shortened, the H-O bonds are lengthened and the oxygen atom of the second sphere water molecules is increased in negative net charge.

This gives rise to the formation of a third hydration sphere and so on. As a result of these interactions, a contraction and a rearrangement in the water structure is established, which is not confined to any natural border lines.

4. Solution Structures

Whereas hydration numbers and hydration structures have been well-established for numerous crystalline hydrates, their estimation in the liquid state is bound to be controversial because of the absence of border lines (see above). The continuous interactions between all molecules throughout the liquid structure lead to a characteristic feature for each individual solution, which always acts as a unity.

Although it is not possible to differentiate clearly between hydration shell and less disturbed water structure, it has been found useful to distinguish hydrated and non-hydrated regions within the solution. The latter contain molecules which are sometimes called "normal" water molecules, because according to the results of spectroscopic investigations they cannot be distinguished from water molecules in the liquid in the absence of solutes.

The statistical results are determined by the enormous number of "normal" water molecules and the differentiations between them are not well pronounced. The statistical results of structural, spectroscopic, thermodynamic and kinetic data of the solution are determined by the vast number of "normal" water molecules, because the properties of water molecules near an interface and around a solute remain "statistically insignificant". (The fundamental difficulties involved in the interpretation of spectroscopic data of liquid solutions will be briefly discussed at the end of the following Chapter and in Chapter 19).

Because of the difficulties in elucidating the results of the actions of solutes on the solution structure (due to the active role of the solutes), most studies in solution chemistry are concerned with the results of the modification of the solutes by the solvent (due to the active role of the solvent and the more passive role of the solute).

References

1.   G. Resch and V. Gutmann, in *Advances in Solution Chemistry*, eds. I. Bertini et al, p. 1, (Plenum Press, New York, 1981).
2.   S. E. Jackson and M. C. R. Symons, *Chem. Phys. Letters* **37** (1976) 551.
3.   I. Olovsson, *J. Chem. Phys.* **49** (1968) 1063.

4. J. O. Lundgren and I. Olovsson, in *The Hydrogen Bond - Recent Developments in Theory and Experiment*, eds. P. Schuster et al, Chapter 10 (North Holland, Publ. Co. Amsterdam, 1976).

5. R. A. Bell, G. G. Christoph, T. R. Fronczek and E. R. Marsh, *Science* **190** (1975) 151.

6. J. P. Hunt, *Metal Ions in Aqueous Solutions*, (Benjamin Inc., New York, Amsterdam, 1963).

7. R. W. Gurney, *Ionic Processes in Solution*, (McGraw-Hill, Publ. Co. London 1953).

8. J. W. Schultz and D. F. Hornig, *J. Phys. Chem.* **65** (1961) 2131.

9. V. Gutmann, *The Donor-Acceptor Approach to Molecular Interactions*, (Plenum Press, New York, 1977).

10. V. Gutmann and G. Resch, *Pure Appl. Chem.* **53** (1981) 1447.

11. G. Resch and V. Gutmann, *Z. physik. Chem. (n. F.)* **121** (1980) 211.

12. V. Gutmann and G. Resch, *Rev. Inorg. Chem.* **2** (1980) 93.

13. H. G. Hertz and C. Raedle, *Ber. Bunsenges. phys. Chem.* **77** (1973) 521.

14. P. Schuster, *Int. J. Quantum Chem.* **3** (1969) 851.

15. P. Russegger, H. Lischka and P. Schuster, *Theor. Chim. Acta* **24** (1972) 191.

CHAPTER 12

WATER AND ALCOHOLS

1. Liquid Alcohols

Primary alcohols containing fewer than 12 carbon atoms are liquid at room temperature. Their liquid range under atmospheric pressure is appreciably wider than that of water (Table 19). Their boiling points increase with higher molecular weight, as do their vapour pressures, densities and viscosities.

Table 19. Liquid range of primary alcohols containing up to 10 carbon atoms.

Alcohol	Melting Point [°C]	Boiling Point [°C]	Liquid Range [°C]
Methanol	- 97	+ 64.7	161.7
Ethanol	- 117	+ 78.4	195.4
1-Propanol	- 126	+ 97.2	223.2
1-Butanol	- 90	+ 117.8	207.8
1-Pentanol	- 78.5	+ 137.8	216.3
1-Hexanol	- 52	+ 156	208
1-Heptanol	- 34	+ 176	210
1-Octanol	- 15	+ 195	210
1-Decanol	+ 6	+ 228	222

The first representatives of this group, methanol, ethanol, 1-propanol as well as t-butanol (m.p. +25.5°C, b.p. +82.4°C) are completely miscible with water. The mutual solubilities between alcohols containing longer carbon chains are limited and the solubilities of alcohols in water are smaller than the solubilities of water in the respective alcohol (Table 20), but it is *impossible to remove the last traces of water from the liquid alcohols.*

From the molecular point of view they exhibit similarities to water. In the liquid state a hydrogen bonded network is established with hydrogen bonding energies for methanol and ethanol nearly as high as those for water. They differ, however, in the number of hydrogen bonds, as four of them are formed by each water molecule but only three by each alcohol molecule. This accounts for the fact that the network in liquid alcohols is looser than in water. This is also reflected in the much wider liquid range of alcohols (Table 19), which extends down to temperatures below 0°C. The greater looseness between the alcohol molecules manifests itself in the low values for vapour pressure, surface tension and thermal conductivity and in high values for heat capacity and gas solubilities.

Table 20. Solubilities between water and n-alcohols at 25°C.

Alcohol	Solubility of Alcohol in Water [% by weight]	Solubility of Water in Alcohol [% by weight]
Methanol	Complete Miscibility	Complete Miscibility
Ethanol	Complete Miscibility	Complete Miscibility
1-Propanol	Complete Miscibility	Complete Miscibility
2-Propanol	Complete Miscibility	Complete Miscibility
1-Butanol	7.45	20.5
t-Butanol	Complete Miscibility	Complete Miscibility
1-Pentanol	2.19	7.46

Both the donor and acceptor properties of alcohols are lower than those of water and they decrease with increasing chain lengths. As the donor and acceptor properties are less developed, the solubility of salts such as sodium chloride is decreased.

Table 21. Donor properties of water and some alcohols, based on the shifts of the spectral bands of oxovanadium (IV) acetylacetonate [1], as well as acceptor properties and solubilities of sodium chloride [2].

Liquid	DN	AN	g NaCl/100 g
Water	42	54	3.57
Methanol	34	41	1.40
Ethanol	30	38	0.065
Propanol	29	37	0.012

Most of the acids in aqueous solutions are also acids in alcoholic solutions and neutralization reactions are found to occur. Due to the well developed donor properties of the alcohols, acceptor halides, such as antimony(V)chloride, titanium(IV)chloride, tin(IV)chloride or aluminium chloride are soluble and solvates may be isolated from their solutions.

Although the donor and acceptor properties of water are better developed than those of the alcohols, iodine is more soluble in liquid alcohols than in water. The iodine molecule is non-polar with poor donor and poor acceptor properties. Its interactions with liquid water are more restricted than those with liquid ethanol, because the stronger donor and acceptor properties of water molecules have been used in the liquid state for the formation of the three-dimensional hydrogen bonded network. The interactions with the iodine molecules are too weak to initiate the structural rearrangement of water. This is accomplished in the presence of potassium iodide, which is ionised with formation of iodide ions, which interact with iodine molecules to give triiodide ions. On the other hand, the network in liquid alcohols is loose enough to allow for the interactions of alcohol molecules with iodine.

The alcohols show regularly changing liquid properties as the aliphatic chain length is increased as does the entropy change on freezing. However, the fractional volume change on freezing shows alternative low (odd number of carbon atoms) and high (even number of carbon atoms) values, namely 8 to 9% for the former and 11 to 13% for the latter [1].

2. Physical Properties of Water - Alcohol Mixtures

Freezing and Boiling Points

Addition of alcohol to water causes a decrease both in freezing and in boiling point. For water-ethanol mixtures a eutectic mixture is formed at 92.4 wt-% ethanol at -123°C and an azeotropic mixture at 95.57 wt-% ethanol at 78.15°C.

Table 22. Freezing and boiling points of water-ethanol mixtures [3].

wt-% ethanol	freezing point [°C]	boiling point [°C]
0	0	100
10	- 4.5	91.45
20	- 10.7	87.15
40	- 27.0	83.10
50	- 37.0	81.90
60	- 45.0	81.0
70	- 53.0	80.20
80	- 64.5	79.35
90	- 109	78.50
92.4 *	- 123	78.24
95.57 **	- 119.3	78.15
100	- 114.5	78.30

*   Eutectic mixture   ** Azeotropic mixture

Table 23. Azeotropic water-alcohol mixtures

Alcohol	wt-%	Azeotropic point [°C]
Ethanol	95.57	78.15
1-Propanol	71.7	87.65
2-Propanol	88.0	80.1
1-Butanol	57.5	92.7
t-Butanol	88.24	79.9
1-Pentanol	45.6	95.8
1-Hexanol	32.8	97.8
1-Octanol	10.0	99.4

Vapour Pressure

Addition of alcohols to water leads to an increase in vapour pressure [4]. Small amounts of alcohol cause greater effects so that the values for the mixtures cannot be calculated according to the theorem of additivities. The differences between calculated and measured values are greatest at an alcohol content of 50 wt-%.

Surface Tension

The surface tension is lowered by addition of alcohols, the lowering being greater as the chain length of the alcohol is increased [5] (Fig. 30).

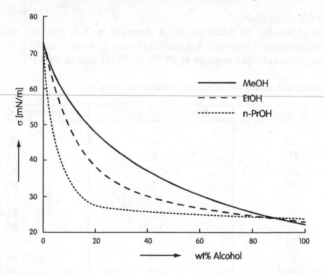

Fig. 30. Surface tension of mixtures of water with methanol, ethanol and propanol.

Density and Volume of Mixing

When alcohol is added to water, the density is decreased and the volume contraction greatest at about equal amounts of alcohol and water [6] (Fig. 31).

Of special interest is the concentration dependence with regard to the influence of alcohol content on the density maximum of pure water at +4°C. Small amounts of ethanol cause a slight increase in the temperature of maximum density (TMD) to +4.3°C at 0.53 mol % ethanol [7].

Table 24. Temperature of maximum density (TMD) for water alcohol mixtures.

Alcohol	TMD [°C]	mol %	wt-%
Ethanol	+ 4.3	0.53	1.34
1-Propanol	+ 4.05	0.24	0.80
2-Propanol	+ 4.18	0.44	1.45
t-Butnaol	+ 4.39	0.43	1.74

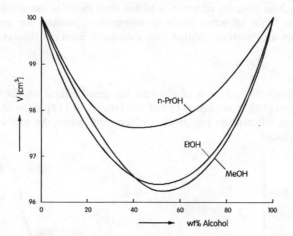

Fig. 31. Volumes of mixing for water - alcohol mixtures at 20 °C.

As the ethanol content is further increased, the density maximum is shifted to lower temperature with decrease in maximum density value. At an ethanol content of 22% the density maximum has disappeared [8] (Fig. 32).

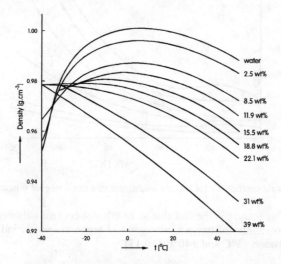

Fig. 32. Temperature dependence of the densities of water - ethanol mixtures.

Molar Heat Capacity

The molar heat capacity of water is higher than those of the alcohols and hence addition of the latter to water leads to decreasing c_p-values for water - alcohol mixtures, which are, however, higher than calculated from the theorem of additivity [9].

Gas Solubilities

Fig. 33 shows that addition of alcohols has small effects on the solubilities of oxygen, nitrogen [10], hydrogen [11] and the rare gases [12] up to about 45 wt-%. The situation is different at higher alcohol content, when the gas solubilities are drastically increased .

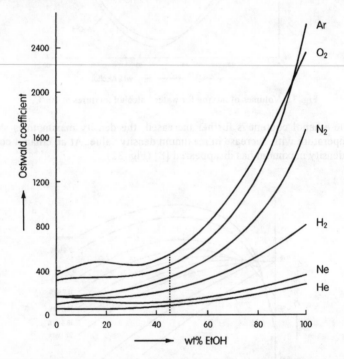

Fig. 33. *Ostwald*-coefficients for the gas solubilities in water - ethanol mixtures at 20°C.

An interesting feature is the fact that at an ethanol content between 35 and 43 % by weight the solubilities of oxygen, nitrogen and argon are nearly independent from temperature between +5°C and +40°C [10,13].

Adiabatic Compressibility

Although the adiabatic compressibility of water is smaller than that of ethanol, it is decreased by addition of ethanol up to about 19 % by weight at 0°C. The decrease is more pronounced the lower the temperature. The temperature dependence between -17°C and +40°C shows that at an ethanol content of about 11 % by weight, corresponding to about 0.05 mol fractions, the compressibility is independent of temperature [14] (Fig. 34).

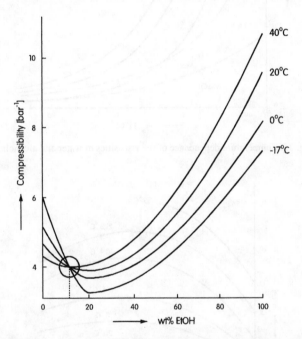

Fig. 34. Adiabatic compressibility of water - ethanol mixtures vs. wt-% ethanol at different temperatures.

Viscosity

The viscosity of liquid methanol is lower than that of liquid water, but that of ethanol slightly greater at +20°C and slightly smaller at +10°C (Fig. 35).

As the ethanol content of water is increased at room temperature, the viscosity is increased and found to pass through a maximum value at 20°C, when the mixture contains about 45 wt-% ethanol [15] (Fig. 36).

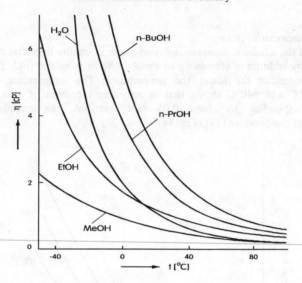

Fig. 35. Temperature dependence of the viscosities of water and alcohols.

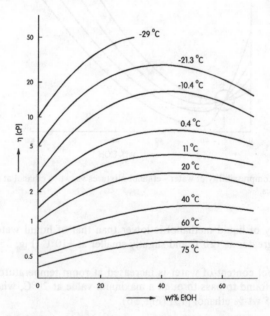

Fig. 36. Viscosity of water - ethanol mixtures *vs*. ethanol content at different temperatures.

3. Physical and Molecular Properties

The changes in physical properties of liquid water due to addition of primary alcohols are similar to those produced by the dissolution of gases in water (Table 25).

Table 25. Actions of gases and of ethanol on the properties of liquid water.

Property	Changes due to addition of gases to water	Changes due to addition of ethanol to water
Vapour Pressure	Increased	Increased
Surface Tension	Decreased	Decreased
Heat Capacity	Increased	Increased
Viscosity	Increased	Increased (up to 45 wt-%)
Density	Decreased	Decreased
Adiabatic Compressibility	Increased	Increased

The hydrophobic properties of alcohols due to the alkyl groups seem to be more important as regarded to the changes in physical properties than their hydrophilic properties (by which they are well integrated in the water structure). Accordingly, the changes in physical properties due to the addition of small amounts of alcohols are frequently more remarkable than those due to addition of larger amounts.

The hydrophobic action of alcohols on water has been well-established by the isolation and structural characterization of clathrates [16-18] of composition $C_2H_5OH.17 H_2O$ with one alcohol molecule entrapped within a void formed by water molecules. These structural elements appear also to be present in the liquid mixtures - at least at a temperature below +30°C [19].

Another point of interest is the comparison of the changes in macroscopic properties due to addition of alcohols to water with those due to the formation of supercooled water (Chapter 6). It may be seen from Table 26 that with the exception of typical surface properties (vapour pressure and surface tension) the bulk properties are changed in the same way.

A solution containing 43 wt-% ethanol freezes at -61°C, but in the solid no structural elements characteristic for ice can be found [17]. Rapid cooling to -170°C gives a glass-like solid material which is more flexible than ice.

It has been noted (Chapter 9), that by addition of alcohols to water the critical micelle concentration (CMC) of surfactants is considerably increased. Only at very low content of alcohol a slight decrease in CMC has been found [21]. Similar effects of alcohol have been reported for the flocculation of hydrophobic sols by electrolytes or by the unfolding of ribonuclease [22]. These findings are in agreement with those of *Ben-Naim* [23] according to which the hydrophobic interaction effect in water is attenuated in alcoholic solutions.

Table 26. Comparison of changes in physical properties of supercooled water by lowering the temperature with liquid water by addition of ethanol at 20°C [20].

Property	Decrease in Temperature causes in Supercooled Water	Addition of Ethanol causes in Liquid Water
Vapour Pressure	Decrease	Increase
Surface Tension	Increase	Decrease
Gas Solubility	Increase	Increase
Heat Capacity	Increase	Increase
Viscosity	Increase	Increase
Density	Decrease	Decrease
Thermal Conductivity	Decrease	Decrease
Absorption of Ultrasound	Increase	Increase
Adiabatic Compressibility	Increase	Increase

A crucial difference between the behaviour of hydrophobic solutes and primary alcohols is provided by the low solubilities of the former and the high solubilities of the latter. This is because

(i) the alcohols are "anchored" and hence well-integrated in the water structure (with appropriate modifications) and

(ii) the structure of the alcohols is gradually changed by the addition of water.

The well-developed hydrophilic properties of short C-chain alcohol molecules has been demonstrated by the results of FTIR-spectroscopy on water containing

Table 27. Comparison of changes in properties by addition of water to alcohol and of hydrophilic solutes to water.

Property	Addition of Water to Alcohol causes	Addition of Hydrophilic Solutes to Water causes
Surface Tension	Increase	Increase
Density	Increase	Increase
Viscosity	Increase	Increase
Gas Solubility	Decrease	Decrease
Heat Capacity	Decrease	Decrease
Compressibility	Decrease	Decrease
Solution Structure	Tightening	Tightening

different amounts of the said alcohols. Addition of water increases the O-H stretching as well as the O-H bending bands, both in width and in intensity, - these changes being typical for hydrophilic solutes.

Addition of short-chain alcohols (up to 1-pentanol) to water causes spectral changes which are similar to those caused by an increase in temperature (Table 28).

Table 28. Spectral changes of water due to added alcohols and due to increase in temperature.

Property	Addition of Alcohol	Increase in Temperature
Transmission	Increased	Increased
O-H-stretch vibration	Narrowed	Narrowed
Absorption of the band at		
2100 cm^{-1}	Decreased	Decreased
Band at 1640 cm^{-1}	Narrowed	Narrowed

Addition of alcohol to water leads to a greater lengthening of the intermolecular O-H bonds, than to shortening of the intramolecular O-H bonds and hence to decreasing O - O distances, as well as to a narrowing of the H-O-H bond angles in agreement with the rules outlined in Chapter 4 [24].

On the other hand, all of these changes occur in opposite directions when water is added to the short-chain alcohols. The spectral changes are outlined in Table 29.

Table 29. Spectral changes for short-chain alcohols due to addition of water and for water due to addition of alcohol.

Property	Changes by Addition of Alcohol to Water	Changes by Addition of Water to Alcohol
Intermolecular O-H bonds	Lengthened	Shortened
Intramolecular O-H bonds	Shortened	Lengthened
H-O-H bond angle	Narrowed	Widened
O - O bond distance	Lengthened	Shortened

From the spectroscopic results the conclusion may be drawn that alcohols added to water behave as *structure breakers*, i.e. as hydrophilic solutes which are well-integrated into the water structure. On the other hand, the changes of macroscopic properties due to addition of alcohol to water are analogous to those produced by addition of *structure makers*, i.e. of hydrophobic solutes.

It is well-known that alcohol molecules contain a hydrophilic as well as a hydrophobic group and hence both of these interactions are expected to take place. It seems, however, impossible to distinguish between the results of these competing influences from the spectroscopic data. These appear to provide the net-effects of both influences on the structural changes. As the net-effects indicate hydrophilic behaviour of the alcohols in water, it may be concluded that the hydrophilic interactions give rise to more pronounced structural changes of water than do the hydrophobic interactions. It might not be correct, however, to ignore or to deny the hydrophobic contributions on grounds of the exclusive consideration of spectroscopic results.

References

1. Y. Marcus, *Introduction to Liquid State Chemistry* (Wiley, New York, 1969)
2. *Gmelin's Handbuch der Anorganischen Chemie* ,Vol. 21/7 (Verlag Chemie, Weinheim, 1973).
3. J. Timmermann, *The Physico-Chemical Constants of Binary Systems in Concentrated Solutions*, Vol. 4 (Interscience, New York, 1960).
4. R. W. Gallant, *Hydrocarbon Process* **45** (1966) 171.
5. F. W. Seemann, *Physikalisch-Technische Bundesanstalt, Braunschweig-Berlin-Mitt.* **91** (2) (1981) 95.
6. P. W. Atkins, *Physikalische Chemie* (Verlag Chemie, Weinheim 1980).
7. G. Wada and S. Umeda, *Bull. Chem. Soc. Japan* **35** (1960) 646.
8. C. M. Sorensen, *J. Chem. Phys.* **79** (1983) 1455.
9. W. S. Knight, *Princeton Univ., Diss. Abstr.* **24** (1963) 993.
10. J. Tokunaga, *J. Chem. Eng. Data* **20** (1975) 41.
11. R. W. Cargill, *J. Chem. Soc., Faraday Trans., I*, **74** (1978) 1441.
12. G. A. Krestov and K. M. Patsatsiya, in H. L. Clever, *Solubility Data Series*, Vol. 1, (IUPAC, Pergamon Press, New York, 1981).
13. A. Ben-Naim, *J. Phys. Chem.* **69** (1965) 3245.
14. O. Conde, J. Teixeira and P. Papon, *J. Chem. Phys.* **76** (1982) 3747.
15. G. W. Euliss, C. M. Sorensen, *J. Chem. Phys.* **80** (1984) 4767.
16. R. J. Speedy, J. A. Ballance and B. D. Cornish, *J. Phys. Chem.* **87** (1983) 325.
17. A. D. Potts and D. W. Davidson, *J. Phys. Chem.* **69** (1965) 996.
18. C. M. Sorensen, *J. Chem. Phys.* **79** (1983) 1455.
19. D. N. Glew, *Nature* **195** (18. Aug. 1962).
20. C. Kuttenberg, G. Scheiber and V. Gutmann, *to be published.*
21. F. Franks, *Water a Comprehensive Treatise*, ed. F. Franks, Vol. 2, p. 88, (Plenum Press, New York 1973).
22. J. F. Brandts and L. Hunt, *J. Amer. Chem. Soc.* **89** (1967) 4826.
23. A. Ben-Naim, *J. Chem. Phys.* **54** (1971) 1387, 3696.
24. V. Gutmann, *The Donor-Acceptor Approach to Molecular Interactions* (Plenum Press, New York, 1977).

CHAPTER 13

CHARACTERIZATION OF NON-AQUEOUS SOLVENTS

1. Historical

Non-aqueous solvents, such as benzene, ether or carbon tetrachloride have been used in organic chemistry for a long time. Other solvents, such as acetonitrile, dimethyl sulfoxide or hexamethylphosphoric triamide have been introduced more recently.

In inorganic chemistry liquid ammonia has been used as a "water-like" solvent since the turn to this century. The self-ionization equilibrium

$$2\,NH_3 \rightleftharpoons NH_4^+ + OH^-$$

and the so-called ammono-system of acids and bases has been established. Liquid sulphur dioxide has been introduced to inorganic chemistry as the first aprotic inorganic solvent by *Walden* [1] and for which *Jander* [2] assumed the self-ionization equilibrium

$$2\,SO_2 \rightleftharpoons SO^{++} + SO_3^{--}$$

in analogy to the self-ionization equilibrium of liquid carbonyl chloride proposed by *Germann* [3],

$$COCl_2 \rightleftharpoons CO^{++} + 2\,Cl^-$$

who considered complex formation between calcium chloride and aluminium chloride in this solvent as a neutralization reaction:

Acid: $\quad 2\,AlCl_3 + COCl_2 \rightleftharpoons CO^{++} + 2\,AlCl_4^-$
Base: $\quad CaCl_2 \rightleftharpoons Ca^{++} + 2\,Cl^-$
Neutralization: $\quad Ca^{++} + 2\,Cl^- + CO^{++} + 2\,AlCl_4^- \rightleftharpoons Ca(AlCl_4)_2 + COCl_2$

This concept was also applied to selenium oxychloride as a solvent [4].

Cady and *Elsey* [5] suggested the general formulation in terms of the so-called *solvent system concept*, according to which an acid is defined as a solute causing an increase in concentration of solvent cations and a base as a solute causing an increase in concentration of solvent anions.

This concept has been applied to describe ionic reactions in various solutions [6], for example, in liquid bromine(III)-fluoride [7]:

Self-ionization:	$2\,BrF_3 \rightleftharpoons BrF_2^+ + BrF_4^-$
Acid:	$BrF_3 + SbF_5 \rightleftharpoons BrF_2^+ + SbF_6^-$
Base:	$KBr + BrF_3 \rightleftharpoons K^+ + BrF_4^-$
Neutralization:	$BrF_2^+SbF_6^- + K^+BrF_4^- \rightleftharpoons KSbF_6 + 2\,BrF_3$

A more general description has been advanced by *Gutmann* and *Lindqvist* [8] by extending the *Brönsted-Lowry* concept. According to this *concept of ionotropism* an acid is either a cation donor or an anion acceptor and a base is either an anion donor or a cation acceptor. This concept provides a bridge to the *Lewis acid-base concept*, proposed in 1923 [9].

Lewis defined a base as an electron pair donor and an acid as an electron pair acceptor. In the long run this concept paved the way for the application of quantum mechanics to chemistry, but it was not accepted by chemists until more than 20 years after its formulation.

The immense value of this concept for coordination chemistry was realized by *Sidgwick* [10]. In organic chemistry, the breakthrough of the *Lewis* ideas was due to *Robinson's* electronic interpretations of organic reactions [11] and its effective extension by *Ingold* [12]. Since then carbon chemists use the term "nucleophile" for a *Lewis* base (donor) and "electrophile" for a *Lewis* acid (acceptor).

In this way the quantum-chemical requirement for the occurrence of charge transfer in the course of any molecular interaction is met and this is not considered by the elementary electrostatic theory.

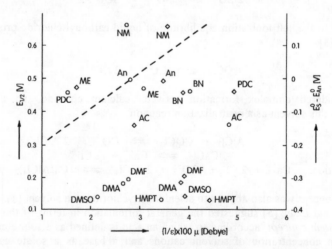

Fig. 37. Half-wave potentials of the system $Co(en)_3^{3+}/Co(en)_3^{2+}$; dipole moments (squares) and dielectric constant (circles). Right hand ordinate standard redox potentials. Dashed line, calc. according to the *Born* equation, E_s standard redox potential in solvent, E_{An}^0 standard redox potential in reference solvent acetonitrile (An).

It can be seen that the parameters of the elementary electrostatic theory are not applicable to non-aqueous solutions. For example, liquid hydrogen cyanide has a higher dielectric constant than water, and yet its solvent properties are extremely limited. Likewise the dipole moment of the acetonitrile molecule is considerably greater than that of the water molecule, but its ionizing properties are inferior to those of liquid water. The limitations of the application of the parameters of the electrostatic theory have been illustrated in a plot of dipole moment and dielectric constant vs. the polarographic half wave potentials for a given redox couple [13] (Fig. 37).

2. Solvent Properties

It has been shown in some detail above that certain aspects of the structural features of liquid water have been established by means of the results of various spectroscopic techniques. These have also been applied to liquid alcohols and other hydrogen bonded solvents.

Unfortunately, the spectra obtained for aprotic solvents are too diffuse to allow reliable conclusions. The difficulties involved in theoretical approaches have also been mentioned in Chapter 5, as all of these are based on certain assumptions on interacting forces, usually on strong repulsive forces at short distances and weak attractive forces at long distances, between solvent molecules considered as hard spheres [14].

None of these approaches considers the fact that, although the liquids are called "non-aqueous solvents", they are bound to contain unremovable traces of water, although the solubility of water in some solvents may be small.

Table 30. Solubility of water in some non-aqueous solvents [15].

Solvent	wt-% of water
Benzene	0.063
Carbon tetrachloride	0.072
Chloroform	0.072
Nitrobenzene	0.246
Diethyl Ether	1.468
Nitromethane	2.090
Acetone	Miscible
Acetonitrile	Miscible

In the absence of structural data and faced with the inapplicability of the parameters of the electrostatic theory, efforts have been restricted to investigations of the influence of the solvent on structures and properties of solutes.

Various empirical solvent parameters have been proposed. *Grunwald* and *Winstein* [16] suggested the Y-values derived from kinetic data. *Kosower* [17] introduced the Z-values, based on the UV-spectra of 1-ethyl-carbomethoxy-pyridinium iodide in the solvents under consideration. The charge transfer energy is adopted as an empirical measure of the solvent polarity and of the ionizing properties of the solvent[1].

   Dimroth and *Reichardt* [21,22] proposed the E_T-values, based on the solvent sensitivity of light absorption of a pyridinium phenol betain.

Kosower's Z-value   Reichardt's E_T-value

These parameters were found applicable in many systems, but appeared useless in a number of other cases (e.g. nucleophilic substitution reactions or cation solvation).

3. The Donor Acceptor Approach

   Based on the extended donor-acceptor approach [18,19] solvent-solvent as well as solute-solvent interactions are described as *Lewis* acid-*Lewis* base interactions. As each solvent has a certain solubility for *Lewis* acids and *Lewis* bases - although the solubilities may be low - it must itself be able to exercise both donor and acceptor functions.

   Since in aprotic solvents these functions are not correlated by a self-ionization equilibrium, it is necessary to use for the characterization of an aprotic solvent not one, but two solvent parameters, namely one for the solvent donor properties and one for the solvent acceptor properties [18,19].

   Lindqvist and *Zackrisson* [20] suggested the ΔH-values for the interactions of the solvent with antimony(V)-chloride or tin(IV)-chloride as a measure of the solvent donor properties. This work has been extended by *Gutmann* and *Wychera* [23], who introduced the *donor number* (DN). This is defined as the molar enthalpy value for the reaction of the solvent with the reference acceptor antimony(V)-chloride in a 10^{-3}M solution of dichloroethane. Due to the existence of LFER's in all

---

[1] The synonymous use of these terms has been misleading, since there is no relationship between solvent polarity (denoting the distribution and polarization of the charges within one solvent molecule) and the ionizing properties (expressing the ability of the solvent as a whole to heterolize more or less developed covalent bonds) [18-20].

of the aprotic solvents, these thermochemically derived values were found linearly related to thermodynamic properties.

For the characterization of the acceptor properties, the *acceptor number* (AN) has been proposed [24] which is derived from the values of the ^{31}P NMR chemical shift in triethylphosphine oxide in the respective pure solvent.

Table 31. Freezing point (f.p.), boiling point (b.p.), donor number (DN) and acceptor number (AN) for various solvents.

Solvent	f.p.[°C]	b.p.[°C]	DN	AN
Acetic acid	+ 17	+ 119	10	53
Acetone (AC)	- 96	+ 56	17	13
Acetonitrile (An)	- 45	+ 82	14	19
Ammonia (liquid)	- 78	- 34	59	-
Benzene	+ 5	+ 80	-	8
Benzonitrile (BN)	- 13	+ 191	12	16
Benzophenone	+ 27	+ 305	16	-
Bromobenzene	- 31	+ 156	10	-
n-Butanol	- 89	+ 118	-	37
t-Butanol	- 24	+ 83	-	27
t-Butylamine	- 68	+ 44	58	-
n-Butyronitrile	- 112	+ 118	17	-
Carbon tetrachloride	- 23	+ 77	-	9
Chloroform	- 64	+ 61	10	23
Chlorobenzene	- 45	+ 132	10	-
1,2-Dichloroethane (DCE)	- 25	+ 84	0*	17
Dichloromethane	- 96	+ 41	10	20
Diethylamine	- 48	+ 56	50	9
Diethyleneglycol dimethylether (diglyme)	- 64	+ 102	24	10
Diethylether	- 116	+ 35	19	4
N,N-Dimethylacetamide (DMA)	- 20	+ 166	28	14
Dimethylformamide (DMF)	- 61	+ 135	27	16
Dimethylsulfoxide (DMSO)	+ 18	+ 189	30	19
Dioxane	- 42	+ 106	19	11
Ethanol	- 115	+ 78	30**	38
Ethanolamine	+ 11	+ 109	50	34
Ethylacetate	- 84	+ 76	17	9
Ethylamine	- 83	+ 17	55	-
Ethylene carbonate	+ 38	+ 236	16	-
Ethylenediamine	+ 9	+ 117	55	21
Ethylene glycol dimethylether (glyme)	- 58	+ 85	20	10
Formamide (FA)	+ 2	+ 105	24	50
Formic acid	+ 8	+ 101	-	84
Hexamethylphosphoric triamide (HMPA)	+ 7	+ 230	39	11

Table 31 continued.

Solvent	f.p.[°C]	b.p.[°C]	DN	AN
Hexane	− 94	+ 69	-	-
Hydrazine	+ 1	+ 113	44	-
Methanol	− 98	+ 65	34**	36
Methyl acetate	− 98	+ 58	17	11
N-Methylformamide (NMF)	− 40	+ 180	27	32
Morpholine	− 7	+ 128	-	18
Nitrobenzene (NB)	− 6	+ 211	4	15
Nitromethane (NM)	− 29	+ 101	3	21
Piperidine	− 9	+ 106	51	-
Phosphorus oxychloride	+ 1	+ 106	12	-
Propanediol-1,2-carbonate (PDC)	− 55	+ 260	15	18
n-Propanol	− 126	+ 97	29**	37
iso-Propanol	− 90	+ 82	-	34
Propionitrile	− 92	+ 97	16	-
Pyridine (PY)	− 42	+ 116	33	14
Selenium oxychloride	− 9	+ 176	12	-
Sulfolane (TMS)	+ 27	+ 285	15	19
Tetrahydrofuran (THF)	− 109	+ 64	20	8
Toluene	− 95	+ 111	-	-
Tributyl phosphate (TBP)	− 79	+ 180	24	10
Triethylamine	− 115	+ 88	61	1
Trifluoroacetic acid	+ 15	+ 73	-	105
Trifluoroethanol	− 44	+ 74	-	53
Trimethylphosphate (TMP)	− 40	+ 197	23	16
Water	0	+ 100	42**	55

* By definition
** For the bulk solvent, based on the shifts of the spectral bands of oxovanadium(IV)-acetylacetonate

Numerical relationships have been found between the acceptor number and the Y-values, Z-values and E_T-values [24] (Fig.38 and Fig.39) and hence these empirical solvent parameters may alternatively be used to characterize solvent acceptor properties [24].

It may be seen from Table 31 that none of the non-aqueous solvents approaches water in equally well-developed donor *and* acceptor properties.

Non-polar solvents, such as benzene or carbon tetrachloride show definitely stronger acceptor properties than polar diethyl ether. The extremely weak acceptor properties and the well-developed donor properties of ethers account for their use as media for reactions involving highly reactive carbanions and for electrophilic substitution reactions.

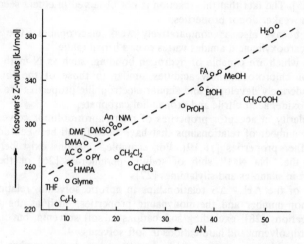

Fig. 38. Relationship between Z-values and acceptor number [24].

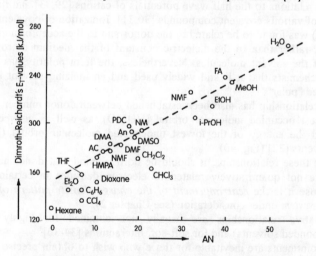

Fig. 39. Relationship between E_T-values and acceptor number [24].

Hexamethylphosphoric triamide is also a poor acceptor solvent, but its donor properties are considerably better developed than those of ethereal solvents. These properties make it suitable as a medium for reactions involving strong and highly reactive bases, such as alkoxides, carbanions etc. In these properties hexamethylphosphoric triamide is similar to liquid ammonia and - although to a lesser extent - it dissolves alkali metals with formation of comparatively stable blue coloured solutions, which appear to contain "free electrons" and strongly solvated metal ions [25]. The fact that this reaction is not observed in ethers seems to be due to their much weaker donor properties.

Ketones behave also as comparatively weak electrophiles. The electrophilic character of carboxylic acid amides varies over a broad range.

Solvents which are capable of hydrogen bonding, such as N-methylformamide, formamide or chloroform show acidities similar to those of the lower aliphatic alcohols. Moderately developed and similar electrophilic properties are exhibited by dimethyl sulfoxide, acetonitrile and propandiol carbonate.

The similarity in acceptor properties of many aprotic donor solvents accounts for the great number of relationships that have been found between donor number and many other properties [18,19]. For example, relations exist between donor number and the ^{23}Na NMR shift of sodium perchlorate [26] or the ^{29}Si NMR chemical shift in silanoles and silylamines [27].

Because of the ΔH - ΔS relationships in aprotic solvents, relationships exist between donor number and thermodynamic properties, such as the transfer free energies of cations [28], excluding soft cations in soft solvents, but including soft cations in hard solvents and hard cations in soft solvents.

The relations between cation solvating power and donor number [28] account also for the relations to the half wave potentials of cations [29,33] and the extent of heterolysis of various covalent compounds [30,31]. Ionization of covalent substrates (heterolysis) was found to be related to the donor and to the acceptor number (push-pull effect), rather than to the dielectric constant of the medium or to the dipole moments of the solvent molecules. Nevertheless, the term polarity has become so familiar to chemists that it is still widely used and an ionizing solvent is therefore usually called "polar".

A fair relationship has also been established between donor number and energy of the highest occupied molecular orbital (HOMO), as well as between acceptor number and the energy of the lowest unoccupied molecular orbital (LUMO) of various solvents [32] (Fig. 40).

Despite these relationships, it should be born in mind that donor and acceptor numbers are not quantitatively related to electron densities in certain molecular areas, because it is the *rearrangement of the charge density pattern of the whole interacting system* under consideration (see Chapter 2).

Quantitative relationships are usually not obtained for highly structured (hydrogen bonded solvents) and for soft-soft interactions [34-36].

Disappointments are inevitable for those who wish to obtain precise predictions on mathematical grounds, as attempted by

(i) the four parameter-equation for the prediction of acid-base reaction enthalpies in the gas phase or in poorly coordinating media by *Drago* [37,38],

(ii) the *Koppel-Palm* concept [39]

(iii) the *Krygowski-Fawcett* equation [40]

(iv) the *Mayer* equation [41]

(v) the *Kamlet-Taft* concept [42,43].

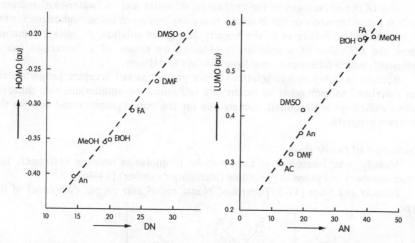

Fig. 40. Relationships between donor number and HOMO, as well as between acceptor number and
        LUMO.

The limitations of all of these approaches have been discussed by *Burger* [34],
who concluded his book on page 256 as follows: "In our experience, the *Gutmann*
donicity scale and the *Gutmann-Mayer* acceptor strength scale can be used most
generally. Accordingly, we propose the determination of these for the
characterization of the solvating powers of solvents (these parameters are already
known for the solvents most widely used in practise)".

It may be added, that the introduction of the donor number and acceptor number
was intended *to provide a guide line for the comparison of different solvents, rather
than for the calculation of quantities. The measured numbers for donor properties
and acceptor properties allow the consideration of differences in numbers referring
to differences in solvent qualities.* Because quality and quantity can never be
completely separated from each other, the differences in numbers are related to
differences in qualities, which, however, cannot be fully reached in this way.

For these reasons, both the donor numbers and the acceptor numbers are meant
in the first place to provide for an orientation in the search for an understanding of
the changes in qualities in the course of interactions in solutions, rather than as an
instrument for the precise calculation of data.

4. Colour Indicators for the Estimation of Donor- and Acceptor Properties

Donor numbers and acceptor numbers have been determined for highly purified
solvents (Table 31). Due to the presence of varying amounts of water or other
solutes, a given solvent or solvent mixture may have donor and acceptor properties
different from those determined for the highly purified solvent.

One of the advantages of the application of water and of "water-like" solvents, such as liquid ammonia or the alcohols, is the application of colour indicators for the estimation of the acidity or of the basicity of a given solution. A quick orientation about the pH-value of a solution is obtained by means of a "universal colour indicator", which differs in colour from pH-unit to pH-unit.

Because in non-protonic solutions donor properties and acceptor properties are not correlated to each other by means of a self-ionization equilibrium, two different colour indicators are required, namely one for the donor property and one for the acceptor property.

Estimation of Donor Properties

Vanadyl acetylacetonate is known to be turquoise in benzene (extremely low donor number) and green in piperidine (high donor number) [44-46].

Fukuda and *Sone* [47-50] prepared planar nickel and copper complexes of the type

and reported their thermochromic and solvatochromic properties [48-50]. The planar nickel complex is transformed by the actions of donor molecules into an octahedral arrangement with simultaneous colour change from red to bluish-green and the observed colour change is due to the existence of the planar and the octahedral forms. These are in equilibrium, which is shifted to the octahedral form as the donor number of the medium is increased [49-52].

$$[\text{Ni}(\beta\text{-dike})(\text{diam})]^+ + 2\,\text{D} \rightleftharpoons [\text{Ni}(\beta\text{-dike})(\text{diam})\text{D}_2]^+$$

square planar	octahedral
diamagnetic	paramagnetic
red	blue or green
Ni - O bonds shorter	Ni - O bonds longer
C = O bonds longer	C = O bonds shorter
C - C bonds shorter	C - C bonds longer

The copper(II) complex appears to be more flexible towards changes in environment, as it is found to undergo continuous structural changes as the donor number is changed. These changes are accompanied by continuous colour changes: violet in nitrobenzene (DN = 4), bluish-violet in dioxane (DN = 15), blue in trimethylphosphate (DN = 23), azure in formamide (DN = 27), turquoise in hexamethylphosphoric triamide (DN = 38) and green in piperidine (DN = 51) [53-

55]. A reference chart has been prepared and a quantitative background given by the relationship between wave length of the absorption maximum in the visible range and the solvent donor number (Fig. 41).

In this way a reference scale has been provided that serves for the quick estimation of the donor number of a given solution or solvent mixture. A spatula-tipful of the Cu-complex is dissolved and the colour compared to that of the reference scale or the wave length of the absorption maximum is determined. The indicator is sufficiently soluble in nearly all solvents and solvent mixtures, but it is decomposed in highly acidic solvents, such as formic acid.

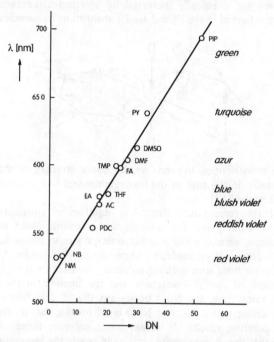

Fig. 41. Relationship between donor number and wave length of the absorption maximum and colour respectively.

Electron-withdrawing substituents at the β-diketone lead to the lengthening of the M-O bonds associated with increase in stability constant of the octahedral species, whereas electron-releasing substituents at the β-diketone lead to shortening of the M-O and M-N bonds and to a decrease in the stability constant. Change of substituents at the diamines leads to changes in the symmetry of the octahedral complexes and to decreasing equilibrium constants in the order of increasing size of the substituents. In both the copper and the nickel complexes, the M-O as well as the M-N bonds are lengthened by coordination of the donor ligands.

Likewise, the C=O bonds of the β-diketone are lengthened, whereas the neighbouring C-C bonds are shortened.

These effects are less pronounced at the respective C-N bonds and the C-C bonds in the diamine. It is interesting to note that electron-releasing substituents at the diamine have the same effect as electron-withdrawing substituents at the β-diketonate. The latter have a more decisive influence than the former. The resulting changes in electron distribution, in bond length and in bond angles are the same as those resulting from the donor influences of the donor solvent. This is found for both, the five- and the six-coordinate species and explains why the stability constants of both of them are drastically increased by electron-withdrawing substituents. These effects are reflected in the IR and far-IR absorption frequencies [56].

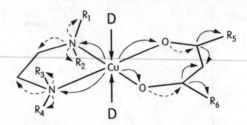

The charge redistribution depends on the donor strength of the solvent donor molecules and leads, in the case of the less symmetrical five-coordinated species to changes in bond angles [57].

Analysis of the temperature dependent equilibrium constants revealed two isokinetic groups, namely one for solvents with medium donor number, such as acetonitrile, acetone, alcohol's and another one for strong donor solvents, such as formamide, N,N-dimethylformamide, pyridine, dimethylsulfoxide. According to the bond angle variation rule, strong donor solvents provoke a strong distortion of the planar arrangement of the β-diketonate and the diamine in the five-coordinate species. At the same time, the Ni-O bonds ant the β-diketonate are weakened. Thus, in strong donor molecules, one Ni-O bond breaks, the β-diketonate rotates and a cis-configuration results. In weak donor solvents these changes are less pronounced, so that the β-diketonate remains in nearly the same position within the molecule and the second donor molecule is simply added to form a trans-isomer.

The decrease in electron density at the electron-releasing substituent of one ligand leads to an increase in electron density at the electron-withdrawing substituent of the other ligand, whereas the electron density near the coordination centre remains nearly invariant.

Estimation of Acceptor Properties

Various complexes are known to change colour by changes in acceptor properties of the medium [58-62]. Among these, bis(phenantroline)iron(II)-dicyanide has been recommended as a colour indicator for the estimation of solvent acceptor properties.

It is sufficiently stable in most solvents, stable towards ligand exchange, readily prepared and even commercially available [63]. It is blue in hexamethylphosphoric triamide (AN=10), bluish-violet in dimethylformamide (AN=16), violet in acetonitrile and dimethylsulfoxide (AN=19), ruby-red in ethanol (AN=38), orange-red in formamide (AN=50) and orange in water (AN=55).

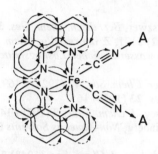

The relationship between wave length of the absorption maximum of the charge transfer band and the acceptor number is shown in Fig. 42 [65].

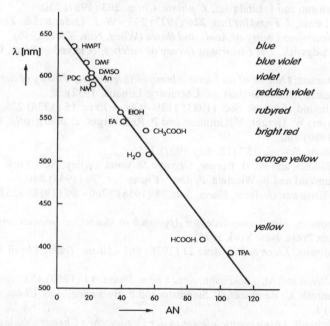

Fig. 42. Relationship between acceptor number and wave length of the absorption maximum as well as colour changes.

In water and acids such as acetic acid, the complex is protonated and the solution yellow. Because the indicator molecules are not deprotonated in aqueous alkaline solutions, the addition of a base leads to a dilution effect rather than to a neutralization: the solution remains yellow.

References

1. P. Walden and M. Centerszwer, *Ber. dtsch. chem. Ges.* **32** (1899) 2862 - *Z. physik. Chem.* **39** (1902) 5131, *Z. anorg. allg. Chem.* **30** (1902) 145, 179.
2. G. Jander, *Die Chemie in wasserähnlichen Lösungsmitteln* (Springer, Berlin 1949).
3. A. F. O. Germann, *J. Amer. Chem. Soc.* **47** (1925) 2469.
4. G. B. L. Smith, *Chem. Rev.* **23** (1938) 165.
5. H. P. Cady and H. M. Elsey, *J. chem. Educ.* **5** (1928) 1425.
6. L. F. Audrieth and J. Kleinberg, *Non-Aqueous Solvents* (Wiley, New York 1953).
7. A. G. Sharpe and H. J. Emeléus, *J. Chem. Soc.* (1948) 2135 - A. A. Woolf and H. J. Emeléus, *J. Chem. Soc.* (1949) 2815, (1950) 1076 - V. Gutmann, *Angew. Chem.* **62** (1950) 312.
8. V. Gutmann and I. Lindqvist, *Z. physik. Chem.* **203** (1954) 250.
9. G. N. Lewis, *J. Franklin Inst.* **226** (1923) 293 - W. F. Luder and S. Zuffanti, *The Electronic Theory of Acids and Bases* (Wiley, New York, 1946).
10. N. V. Sidgwick, *The Electronic Theory of Valency*, (Clarendon Press, Oxford 1927).
11. R. Robinson, *Outline of an Electrochemical (Electronic) Theory of the Course of Organic Reactions* (Inst. of Chemistry, London, 1932).
12. C. K. Ingold, *J. Chem. Soc.* (1933) 1120 - *Chem. Revs.* **15** (1934) 225.
13. U. Mayer, W. Gerger, V. Gutmann and P. Rechberger, *Z. anorg. allg. Chem.* **464** (1980) 200.
14. B. Widom, *Science* **157** (28. July 1967) 375.
15. J. A. Riddick and W. B. Burger, *Organic Solvents*, (Wiley, New York, 1970).
16. E. Grunwald and S. Winstein, *J. Amer. Chem. Soc.* **70** (1948) 846.
17. E. M. Kosower, *J. Amer. Chem. Soc.* **78** (1956) 5700 - **80** (1958) 3253, 3261, 3267.
18. V. Gutmann, *The Donor-Acceptor Approach to Molecular Interactions* (Plenum Press, New York 1978).
19. V. Gutmann, *Electrochim. Acta* **21** (1976) 661 - *Chem. Techn.* (April 1977) 255.
20. I. Lindqvist and M. Zackrisson, *Acta Chem. Scand.* **14** (1960) 453.
21. K. Dimroth, C. Reichardt, T. Siepmann and F. Bohlmann, *Ann. Chem.* **661** (1963) 1.
22. C. Reichardt, *Lösungsmitteleffekte in der Organischen Chemie* (Verlag Chemie Weinheim 1969).
23. V. Gutmann and E. Wychera, *Inorg. Nucl. Chem. Letters* **2** (1966) 257.

24. U. Mayer, V. Gutmann and W. Gerger, *Monatsh. Chem.* **106** (1975) 1235.
25. H. Normant, *Angew. Chem.* **79** (1967) 1029.
26. R. H. Erlich and A. I. Popov, *J. Amer. Chem. Soc.* **93** (1971) 5620.
27. E. A. Williams, J. Cargioli and R. W. Larochelle, *J. Organomet. Chem.* **108** (1976) 153.
28. A. J. Parker, U. Mayer, R. Schmid and V. Gutmann, *J. Org. Chem.* **43** (1978) 1843.
29. V. Gutmann and R. Schmid, *Monatsh. Chem.* **100** (1969) 2133 - V. Gutmann, *Structure and Bonding* **15** (1973) 141.
30. V. Gutmann, *Angew. Chem.* **82** (1970) 858, Int. Ed. **9** (1970) 843.
31. V. Gutmann and U. Mayer, *Monatsh. Chem.* **100** (1969) 2048.
32. A. Sabatino, G. LaManna and I. Paolini, *J. Phys. Chem.* **84** (1980) 2641.
33. G. Gritzner, K. Danksagmüller and V. Gutmann, *J. Elektroanal. Chem.* **90** (1978) 203.
34. K. Burger, *Solvation, Ionic and Complex Formation Reactions in Non-Aqueous Solvents*, (Akademiai Kiado, Budapest 1983).
35. W. B. Jensen, *The Lewis Acid Base Concepts - an Overview* (Wiley, New York 1980).
36. R. Schmid and V. N. Sapunov, *Non-Formal Kinetics* (Verlag Chemie Weinheim, 1982).
37. R. S. Drago and B. Wayland, *J. Amer. Chem. Soc.* **87** (1965) 3571.
38. R. S. Drago, *Structure and Bonding* **15** (1973) 73.
39. A. I. Koppel and U. A. Palm, in *Advances in Linear Free Energy Relationships* Chapter 5, ed. N. Chapmann and S. Shorter, (Plenum Press, New York 1972).
40. T. M. Krygowski and W. R. Fawcett, *J. Amer. Chem. Soc.* **97** (1975) 2143.
41. U. Mayer, *Pure Appl. Chem.* **51** (1979) 1697.
42. M. J. Kamlet and R. W. Taft, *J. Amer. Chem. Soc.* **98** (1976) 377.
43. R. W. Taft, N. J. Pienta, M. Kamlet and E. M. Arnett, *J. Org. Chem.* **46** (1981) 661.
44. R. I. Carlin and F. A. Walker, *J. Amer. Chem. Soc.* **87** (1965) 2128.
45. V. Gutmann and U. Mayer, *Monatsh. Chem.* **99** (1968) 1383.
46. A. Urbanczyk and M. K. Kalinowski, *Monatsh. Chem.* **114** (9183) 1311.
47. Y. Fukuda and K. Sone, *J. Inorg. Nucl. Chem.* **37** (1975) 455.
48. Y. Fukuda and K. Sone, *J. Inorg. Nucl. Chem.* **34** (1972) 315.
49. K. Sone and Y. Fukuda, *Studies in Phys. and Theor. Chem.* **27** (1982) 251.
50. K. Sone and Y. Fukuda, *Inorganic Thermochromism*, Vol. 10, Inorganic Chemistry Concepts (Springer, Heidelberg, New York 1987).
51. V. Gutmann, *Chim. Oggi* **7** (June 1989) 23.
52. W. Linert, V. Gutmann, B. Pouresmaeil and R. F. Jameson, *Electrochim. Acta* **17** (1988) 25.
53. R. W. Soukup, *Chemie in unserer Zeit* **17** (1983) 129.
54. R. W. Soukup and R. Schmid, *J. Chem. Educ.* **62** (1985) 459.
55. V. Gutmann, in *Metal Complexes in Solution*, ed. E. A. Jenne et al, p. 205, (Piccin Press, Padova 1986).

56. W. Linert, V. Gutmann, A. Taha and E. M. Salah, *Rev. Roum. Chim.* **36** (1991) 507.

57. A. Taha, V. Gutmann and W. Linert, *Monatsh. Chem.* **122** (1991) 327.

58. J. Bjerrum, A. W. Adamson and O. Bostrup, *Acta Chem. Scand.* **10** (1956) 329.

59. J. Burgess, *J. Organometal. Chem.* **19** (1969) 218.

60. J. Burgess, *Spectrochim. Acta* **26A** (1970) 1369.

61. J. Burgess, *Spectrochim. Acta* **26A** (1970) 1957.

62. N. Sanders and P. Day, *J. Chem. Soc.* A (1969) 2303.

63. Alfa Ventron 12 189.

64. D. F. Shriver and J. Posner, *J. Amer. Chem. Soc.* **88** (1966) 1672.

65. V. Gutmann and G. Resch, *Monatsh. Chem.* **119** (1988) 1251.

66. W. Linert, V. Gutmann, O. Baumgartner, G. Wiesinger and H. Kirchmayr, *Inorg. Chim. Acta* **74** (1983) 123.

CHAPTER 14

SOLVATION IN NON-AQUEOUS SOLVENTS

1. Cation Solvation

As the donor number is obtained by calorimetric measurements, it is related to the enthalpies of solvation of cations. Because of the existence of free linear energy relationships for donor solvent-acceptor interactions, linear relationships exist between the enthalpic terms of donor numbers and thermodynamic quantities. Fig. 43 shows the linear relationship between the free enthalpies of transfer from the reference solvent acetonitrile into the solvent under consideration for potassium ions and the donor number of the solvent S [1]. These data are based on the assumption of negligible liquid junction potentials in the cell K(Hg)KClO$_4$ in S/Bu$_4$NClO$_4$ in CH$_3$CN/KClO$_4$ in CH$_3$CN/K(Hg).

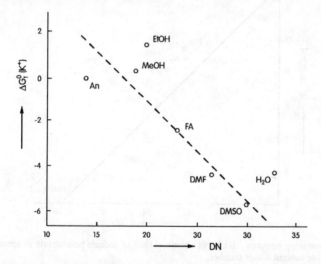

Fig. 43. Correlation between the free enthalpies of transfer of the potassium ion in various solvents obtained from EMF measurements using a salt bridge of 0,1 M Bu$_4$NClO$_4$ and the solvent donor number [2].

According to this relationship, the use of the donor number provides another extrathermodynamic approach to obtain thermodynamic data for ionic species.

According to the relationship

$$-\Delta G^0 = nFE^0$$

linear relationships are found between the standard redox potential of the hydrogen electrode in various solvents and the solvent donor number [3], as well as between the polarographic half wave potentials of many cations, such as Na^+, K^+, Tl^+, Zn^{2+}, Cd^{2+}, Sm^{3+}, Eu^{3+} and the solvent donor number [4-6]. The redox potentials are shifted to more negative values as the solvent donor number is increased. The solvation of the higher-valiant form of such hard cations is usually stronger than that of the lower-valiant forms and hence these differences are reflected in shifts of the redox potentials to more negative values.

These shifts are expressions of the decrease in positive net charge of the cations due to solvation (which is greater for a cation of higher charge than for that of lower charge) and this is in agreement with the description of hydration phenomena [6].

The decrease in positive net charge of the sodium ion by solvation is well demonstrated by the linear relationship between the ^{23}Na NMR chemical shift in solutions of sodium perchlorate or sodium tetraphenylborate and the solvent donor number [7-9] (Fig. 44), whereas no relations are found with dipole moment or dielectric constant of the solvent.

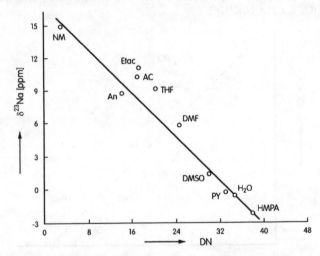

Fig. 44. Relationship between ^{23}Na NMR chemical shift of sodium perchlorate in aprotic solvents and the solvent donor number.

Analogous relationships have also been found for complex cations. Due to outer-sphere coordination of the trisethylenediamine cobalt(III) ion by solvent molecules, the chemical shifts in the ^{59}Co NMR spectra are found at higher field as the solvent donor number is increased [10]. The decrease in positive net charge at the coordination centre is also reflected in the relationship between the polarographic half-wave potential and the ^{59}Co NMR chemical shift [11].

Accordingly one might expect outer-sphere coordination by solvent molecules, although to a small extent, in *trisphenantroline iron complexes* with subsequent decrease in netcharge at the coordination centre and appropriate changes in redox properties. However, the redox properties of the system Fe(tmphen)₃(II/III) (tmphen = 3,4,7,8,tetramethyl 1,10 phenantroline) are nearly independent of the solvent. It has been suggested that nucleophilic attack of solvent molecules occurs at the coordination centre [12,13], rather than an outer-sphere effect.

The complex is red and diamagnetic in the reduced state and blue and paramagnetic in the oxidised state. It is therefore assumed that the electron density around the iron nucleus is greater in the reduced than in the oxidised species. The diamagnetism of the red form is regarded as due to the pairwise arrangement of six 3d-electrons and the paramagnetism of the blue form as due to the presence of one unpaired electron out of only five 3d-electrons.

$$[Fe(phen)_3]^{2+} \rightleftharpoons [Fe(phen)_3]^{3+} + e$$

$$\begin{matrix} \text{red} & \text{blue} \\ 3d^6 & 3d^5 \\ \text{diamagnetic} & \text{paramagnetic} \end{matrix}$$

From these considerations one might expect different electron densities around the iron nucleus and hence different Fe-N-distances in the reduced and in the oxidized form. However, X-ray structural analysis does not show any such differences in electron densities around the iron nuclei. Even by substitutions at the phenantroline rings, the *Mössbauer* isomer shifts remain nearly invariant [14]. A close examination of the *Mössbauer* results reveals that the small differences found in the isomer shifts indicate *an even greater electron density in the oxidized that in the reduced form* [15].

The results obtained by quantum-chemical CNDO/2 calculations point in the same direction [15]: The partial charges at the iron atom are close to zero both in the reduced and in the oxidized complex ion. According to the results of the calculations [15], the charge differences between oxidised and reduced species are most pronounced at the terminating C-H groups: the peripheric H-atoms have higher positive net charges in the oxidized than in the reduced complex ion (Fig. 45) in agreement with their different behaviour towards solvents [16].

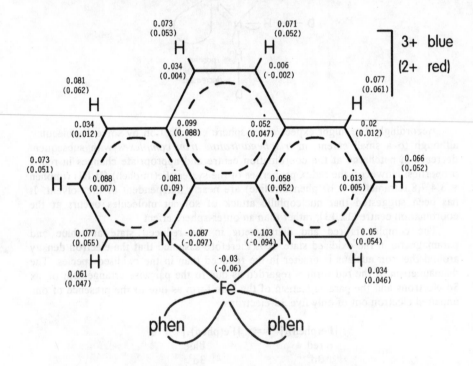

Fig. 45. Results of quantum-chemical CNDO/2 calculations for tris-phenanthroline iron complexes in the reduced form (in brackets) and in the oxidized form [15].

Almost no changes in electron densities occur by substituent and solvent effects. An isokinetic relationship is found for the rate coefficients for the reduction of $[Fe(phenX)_3]^{3+}$ (with different substituents X) by hydrated Fe^{2+} [15] (Fig. 46).

The rate coefficients for the electron transfer reactions in the course of the redox reaction of (3,4,7,8-tetramethyl)-phenanthroline iron (II) by FeL_6^{3+} for different solvent ligands L are found to be related to the donor number [17]. For L = DMSO the rate of reaction is found independent of changes in temperature and the same as extrapolated for the isokinetic temperature [17] (Fig. 47).

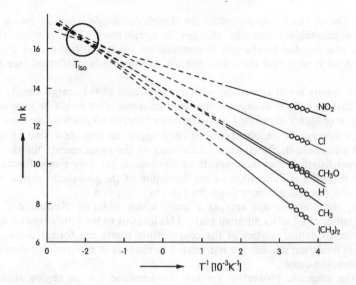

Fig. 46. Arrhenius plot for the rate coefficients for the reaction of [Fe(phen X)₃]³⁺ by hydrated Fe²⁺ in dependence of the substituent X in 5-position.

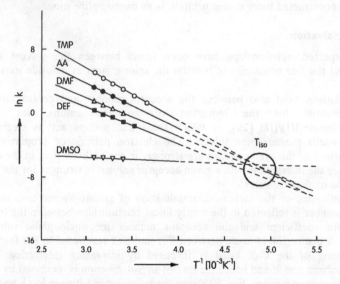

Fig. 47. Arrhenius plot for the rate coefficients for the decomposition of [FeL₅³⁺. Fe(tmphen)₃]²⁺ (tmphen = 3,4,7,8,-tetramethyl-1,10 phenantroline, L = trimethylphosphate TMP, acetamide AA, dimethylformamide DMF, diethylformamide DEF and dimethylsulfoxide DMSO).

This shows that, unexpectedly, the dimethylsulfoxide solution has a highly developed adaptability towards changes in temperature. This requires that all energetic changes due to changes in temperature are redistributed over the whole solution in such ways that the solute species remains nearly unaffected (see section 20.3).

These results are in agreement with the invariance of the charge density around the iron nucleus and the changes at the terminal atoms. *This would be impossible in the absence of highly developed cooperativities between all parts of the system.* The π-electron systems of the phenantroline rings appear to provide a kind of a highly balanced electronic "buffer-system". According to the requirements for the charge density redistribution they are capable of developing not only donor functions (as expected from the consideration of the formation of the so-called "charge transfer complexes"), but also acceptor functions [13,15].

Thus, the complex ion acts as a unity which is highly flexible due to the cooperativities of all of its different parts. This leads us to the following conclusions:

(i) the oxidation number of the coordination centre has formal character and changes in oxidation number are not related to changes in electron densities around the coordination centre,

(ii) the magnetic properties are not characteristic for the region around the coordination centre, but rather for the whole of the complex ion,

(iii) the description of electron densities in terms of molecular orbitals, artificially constructed from atomic orbitals, is an oversimplification.

2. Anion Solvation

As expected, relationships have been found between the solvent acceptor number and the free enthalpies of transfer for anions, such as chloride ions [18,19] (Fig. 48).

Correlations exist also between the acceptor number and polarographic half wave potential for the reduction of complex anions, for example hexacyanoferrate(III)/(II) [20], in which the cyano groups act as electron pair donors towards the acceptor solvent. The electron pair donor properties at the nitrogen atoms of the cyano groups are stronger in the reduced than in the oxidized state. Hence the interaction with a given acceptor solvent is stronger for the reduced than for the oxidized species.

The influence of the increasing stabilization of anions by increase in solvent acceptor number is reflected in the nearly linear relationships between the logarithm of the rate coefficient and the acceptor number for nucleophilic substitution reactions. A distinction is made between S_N1 and S_N2 reactions. In the former case the loosening of the C-X bond is favoured by increasing interaction between acceptor solvent and X and hence the rate of an S_N1 reaction is increased by increase in solvent acceptor number. Fig. 50 shows the relationship between log k and solvent acceptor number [19]:

$$R_3C - X + n\,Acc \rightleftharpoons [R_3C]^+ + [X\,(Acc)_n]^-$$

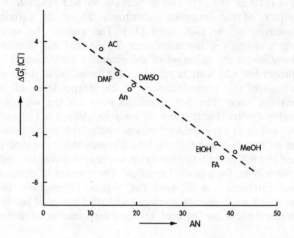

Fig. 48. Relationship between the free enthalpies of transfer for chloride ions and the solvent
acceptor number.

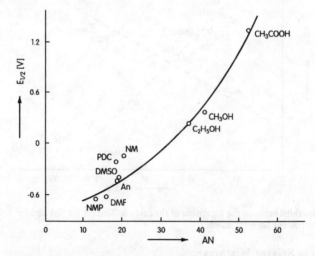

Fig. 49. Relationship between polarographic half wave potential for the redox reaction hexacyano-
ferrate(III)/(II) in different solvents and the solvent acceptor number [20].

On the other hand the rate coefficient of an S_N2 reaction is decreased as the
acceptor number is increased. For example, a nearly linear relationship has been
found between log k and the solvent acceptor number for the solvolysis of p-

methoxyneophyl tosylate [19,21]. This is because an S_N2 reaction is preceded by nucleophilic attack of the incoming substituent Y^- at the carbon atom with subsequent loosening of the C-X bond [19]. The greater the solvent acceptor number, the more strongly is the substituent Y^- solvated and decreased in donor properties. The stronger the solvation of the incoming substituent Y^-, the smaller are its donor properties, which are required for the formation of the transition state.

The advantage of this general concept is its independence of interpretations about the transition state. The S_N1 transition state for the solvolysis of RX is indisputably highly dipolar, *but it is not an ion pair.* Also it is certainly not solvent separated ions, and it is very unlike "normal" polarized organic molecules [19]. Many chemists do not wish to be drawn into disputes about the type of interaction between solvent and RX in the transition state, but they are very interested in finding the most suitable solvent for a desired reaction. The concept of donor interaction at R and acceptor interaction at X does not require assumptions about bonding interpretations, which are frequently confused with the "nature of bonding", nor does it require a statement about the "nature", i.e. the interpretation, of the interactions.

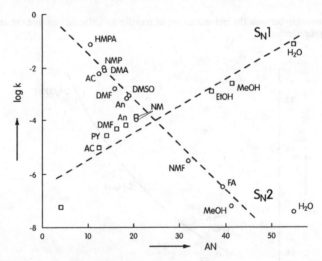

Fig. 50. Relationships between the logarithm of the rate coefficients for S_N1 and for S_N2 reactions and the solvent acceptor number.

3. Solvation in Solvent Mixtures

Preferential Solvation

It is impossible to predict solvation numbers from the theorem of additivity of the solvation properties of the component solvents. For example, in a mixture of 10% water (DN=42) and 90% methanol (DN=34) nickel(II) is solvated in the first solvation sphere by one methanol molecule and five water molecules [24].

In a solvent mixture a given cation is preferentially solvated by the molecules of the stronger donor solvent and a given anion by the molecules of the stronger acceptor solvent. The so-called iso-solvation point has been introduced in order to indicate the composition of the mixture in which the ion under consideration is solvated by an equal number of molecules from each of the components of the solvent mixture.

According to the results of ^{23}Na NMR measurements in different solvent mixtures the iso-solvation points are found at about 0,05 mole-% HMPA (DN=39) in a mixture with nitromethane (DN=3). The iso-solvation points marked in Fig. 51 have been constructed as corresponding to the mean values of the donor numbers of the components, i.e. for the above mentioned system for a DN=21.

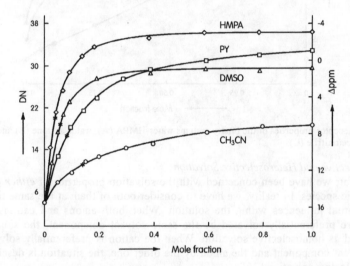

Fig. 51. Variation of the chemical shift of the ^{23}Na NMR resonance of sodium perchlorate
        solutions in binary solvent mixtures of nitromethane with the solvent indicated in the
        plot. Iso-solvation points are indicated by a small star.

The acceptor properties of solvent mixtures usually follow the same pattern, for example in the system water-acetonitrile [25]. A different behaviour has been found for the system water-hexamethylphosphoric triamide, where the weaker acceptor solvent HMPA (AN=11) has a stronger influence on the acceptor properties of the mixture than the stronger acceptor solvent water (AN=55) (Fig. 52). *This is because each solvent mixture represents a system of its own with characteristic structural and functional features.*

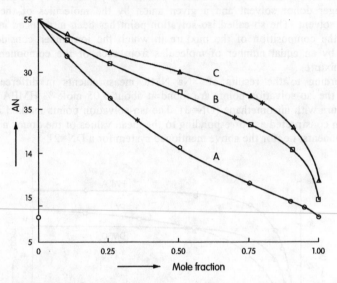

Fig. 52. Acceptor properties of the binary systems water-HMPA (A), water-acetone (B) and water-acetonitrile (C).

Homoselective and Heteroselective Solvation

So far, we have been concerned with the solvation properties of *either* cationic *or* anionic species. In reality, we have to consider both of them at the same time and their mutual influences within the solution. When both anions and cations of the solute are preferentially solvated by the same solvent component, the situation is described as homoselective solvation. When the cation is preferentially solvated by one solvent component and the anion by the other one, the situation is described as heteroselective solvation [26].

Fig. 53 shows heterosolvation of silver nitrate in water-acetonitrile mixtures, the silver ion being preferentially solvated by acetonitrile molecules and the nitrate ion preferentially solvated by water molecules. Fig. 53 shows the behaviour of calcium chloride in water-methanol mixtures as an example for homoselective solvation as both ions are preferentially solvated by water molecules. In systems with heteroselective solvation the solute solubility is higher in the solvent mixture than in either of the pure component solvents [28,29].

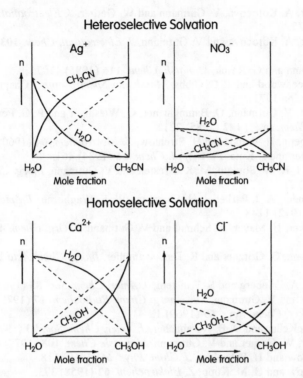

Fig. 53. Heteroselective solvation of silver nitrate in water-acetonitrile mixtures and homoselective solvation of calcium chloride in water-methanol mixtures [27].

References

1. V. Gutmann, G. Resch and W. Linert, *Coord. Chem. Revs.* **43** (1982) 133.
2. D. A. Owensby, A. J. Parker and J. W. Diggle, *J. Amer. Chem. Soc.* **96** (1974) 2682.
3. D. Bauer and A. Foucault, *J. Electroanal. Chem.* **67** (1976) 19.
4. O. Duschek and V. Gutmann, *Monatsh. Chem.* **104** (1973) 990.
5. V. Gutmann and G. Peychal-Heiling, *Monatsh. Chem.* **100** (1969) 1423.
6. V. Gutmann, *The Donor-Acceptor Approach to Molecular Interactions* (Plenum, New York, 1977).
7. R. H. Erlich, E. Roach and A. I. Popov, *J. Amer. Chem. Soc.* **92** (1970) 4989.
8. R. H. Erlich and A. I. Popov, *J. Amer. Chem. Soc.* **93** (1971) 5620.
9. A. I. Popov, *Pure Appl. Chem.* **41** (1975) 275.
10. G. Gonzalez, U. Mayer and V. Gutmann, *Inorg. Nucl. Chem. Letters* **15** (1979) 155.

156   Lecture Notes on Solution Chemistry

11. U. Mayer, A. Kotocová, V. Gutmann and W. Gerger, *J. Electroanal. Chem.* **100** (1979) 875.
12. U. Mayer, A. Kotocová and V. Gutmann, *J. Electroanal. Chem.* **103** (1979) 409.
13. V. Gutmann and G. Resch, *Monatsh. Chem.* **116** (1985) 1107.
14. N. N. Greenwood and T. C. Gibbs, *Mössbauer Spectroscopy,* (Chapman and Hall, London 1971).
15. W. Linert, V. Gutmann, O. Baumgartner, G. Wiesinger and P. G. Perkins, *Z. physik. Chem.* (n.F.) **142** (1985) 221.
16. M. H. Koepp, H. Wendt and H. Strehlow, *Z. Elektrochem.* **64** (1960) 483 - G. Gritzner and J. Kuta, *Pure Appl. Chem.* **56** (1984) 461.
17. R. Schmid, R. W. Soukup, M. K. Aresteh and V. Gutmann, *Inorg. Chim Acta* **73** (1983) 21.
18. R. Alexander, A. J. Parker, J. H. Sharp and W. E. Waghorne, *J. Amer. Chem. Soc.* **94** (1972) 1148.
19. A. J. Parker, U. Mayer, R. Schmid and V. Gutmann, *J. Org. Chem.* **43** (1978) 1843.
20. V. Gutmann, G. Gritzner and K. Danksagmüller, *Inorg. Chim. Acta* **17** (1976) 81.
21. S. Smith, A. Fainberg and S. Winstein, *J. Amer. Chem. Soc.* **83** (1961) 618.
22. U. Mayer and V. Gutmann, *Adv. Inorg. Chem. Radiochem.* **17** (1975) 189.
23. A. J. Parker, *Chem. Revs.* **69** (1969) 1.
24. W. J. MacKellar and D. B. Rorabacher, *J. Amer. Chem. Soc.* **93** (1971) 4379.
25. U. Mayer, W. Gerger and V. Gutmann, *Monatsh. Chem.* **108** (1977) 489.
26. H. Strehlow and H. Schneider, *J. Chim. Phys.* **66** (1969) 118.
27. H. Strehlow and H. M. Köpp, *Z. Elektrochem.* **62** (1958) 373.
28. W. L. Driessen and W. L. Groeneveld, *Rec. Trav. Chim.* **87** (1968) 786
29. W. L. Driessen and W. L. Groeneveld, *Rec. Trav. Chim.* **88** (1969) 620.

CHAPTER 15

IONIZATION AND ASSOCIATION IN NON-AQUEOUS SOLUTIONS

1. Ionization

Until recently the choice of a solvent for a particular reaction has been made empirically [1]. For example, diethyl ether has been used for a long time in order to carry out various substitution reactions in organic chemistry. Although ionic intermediates appear to be involved in many substitution reactions, their formation in diethyl ether could not be explained by elementary electrostatic theory.

It has been convincingly demonstrated more than 25 years ago, that even in water, ionization of a covalent substrate is dependent on its strongly developed coordinating properties rather than by its dipole moment or its dielectric constant [2,3]. It has been shown that the ionizing properties of a solvent are related both to its donor number and to its acceptor number, because the solvation enthalpies for a given cation and for a given anion are related to the donor number and to the acceptor number, respectively of the solvent [2,4].

The influence of the donor number has been established by conductometric titrations of trimethyltin iodide dissolved in nitrobenzene with solvents of different donor number, but similar acceptor number [3] (Fig. 54).

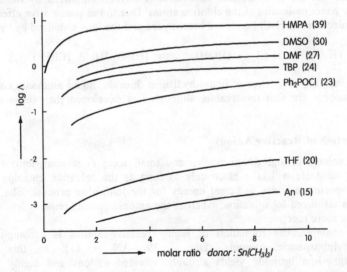

Fig. 54. Conductometric titrations of solutions of trimethyltin iodide in nitrobenzene (dielectric constant of 36) with donor solvents (donor numbers given in brackets).

Nitrobenzene has a reasonable dielectric constant of 36, but small donor number and small acceptor number, so that trimethyltin iodide is practically unionised in the solution. This solution becomes conducting as donor solvents are added, the increase in conductivity being related to the amount of donor present and to its donor number: ionization is accomplished by stabilization of solute cations by solvation, whereas the iodide ions are less solvated [3]:

$$n\,D \;+\; (CH_3)_3SnI \;\; \rightleftharpoons \;\; [D_nSn(CH_3)_3]^+ \;+\; I^-$$

The higher the dielectric constant of the medium, the more is the equilibrium shifted to the right side of the equation. In the present case an equilibrium is established between the ionized species indicated in the equation and so-called ion pairs (see section 4).

A solvent of strong donor properties *and* of strong acceptor properties is an excellent ionizing solvent because of the stabilization of both cationic and anionic species by solvation. Many ionic and covalent compounds give conducting solutions in water due to the formation of hydrated ions, ionic equilibria are established and ionic reactions readily take place.

It has been emphasized that water is outstanding in its ionizing properties, both for ionic and covalent compounds. According to the donor-acceptor approach, ionization of hydrogen chloride is explained by the donor attack of water molecules at the hydrogen atoms of the hydrogen chloride molecules with a considerable shift in electron density from the hydrogen atoms towards the chlorine atoms and polarization of the H-Cl bonds. This trend is further supported by the acceptor attack by water molecules at the chlorine atoms. Due to the cooperative effects of all water molecules the H-Cl bonds are heterolyzed and the ions stabilized by hydration:

$$(H_2O)_n \;\rightarrow\; H\text{-}Cl \;\rightarrow\; (HOH)_m \;\; \rightleftharpoons \;\; [(H_2O)_nH]^+ \;+\; [Cl(HOH)_m]^-$$

The ionizing properties of liquid hydrogen fluoride, liquid ammonia and of the lower alcohols are also remarkable, although less developed than those of liquid water.

2. Formation of Reactive Anions

In a solvent of high donor number and small acceptor number, ionization of a covalent substrate will take place only as long as the solvation enthalpy for the cationic species provides sufficient energy for the ionization process. The cationic species is stabilized by solvation, whereas the anionic species remains less solvated and hence more reactive.

For example, the formation of highly reactive anions is accomplished in hexamethylphosphoric triamide (DN = 39, AN = 11). In this solvent benzylmagnesium bromide yields strongly solvated cations and highly reactive carbanions [5]:

$$C_6H_5\text{-}CH_2MgBr \ + \ n \ HMPA \ \rightleftharpoons \ [(HMPA)_nMgBr]^+ \ + \ [C_6H_5\text{-}CH_2]^-$$

Hexamethylphosphoric triamide is an excellent medium for many nucleophilic substitution reactions, for which diethyl ether is frequently used as a solvent. Its usefulness is due to its reasonably high donor number (DN = 19) and its even smaller acceptor number (AN = 4) than that of HMPA. This is the reason why diethyl ether is a useful solvent for the performance of *Grignard*-type reactions, where a suitable polarization of the Mg-C bond must be achieved with sufficient increase in negative fractional charge for the partial development of a carbanion with well developed nucleophilic properties.

The role of the polarizability of the metal-carbon bond is seen from the comparison of the behaviour of lithium alkyl and of sodium alkyl compounds. Methyl lithium has the molecular structure of a tetrahedron with the methyl groups on each of its faces, so that each lithium atom is coordinated by three methyl groups [6]. This structure is known to prevail in benzene solution and coordination by base molecules occurs to each of the lithium positions [7] with subsequent polarization of the Li-C bonds. The extent of the metal-carbon bond polarization and that of ionization depends on the donor properties of the solvent with low acceptor properties and increases in the following order: benzene < diethyl ether < tetrahydrofurane < pyridine < hexamethylphosphoric triamide [8,9].

On the other hand, ethyl sodium has a layer structure [10]. The bonds are believed to be more ionic than in the lithium compound and this is in agreement with the poor solubility of the sodium compound in most organic solvents.

The reactivity of a given organolithium compound depends on the solvent donor properties. Aliphatic nitriles are deprotonated by organolithium compounds in diethyl ether [11], whereas no reaction takes place in petroleum ether. The carbanions produced in diethyl ether attack the methyl groups of the nitriles with deprotonation. The constitutional contributions to the stabilities of carbanions have been reviewed by *Cram* [12].

The acidity of metal ions and the reactivities of organometallic compounds decrease in the order [9]: Be > Mg > Li > Ca > Na > Sr > Ba > K > Rb > Cs > R$_4$N. This behaviour is also known for inorganic salts: in HMPA lithium iodide has a higher conductivity and is therefore more extensively ionized than tetraethylammonium iodide in the same solvent [13].

Ethereal solvents are applied in many organometallic reactions as well as in inorganic synthesis. An example is the formation of highly reactive [AlH$_4$]$^-$ ions in the course of the preparation of diborane:

$$n \ R_2O \ + \ LiAlH_4 \ \rightleftharpoons \ [(R_2O)_nLi]^+ \ + \ [AlH_4]^-$$

$$4 \ BCl_3 \ + \ 3 \ [AlH_4]^- \ \rightleftharpoons \ 2 \ B_2H_6 \ + \ 3 \ [AlCl_4]^-$$

The rates of nucleophilic substitution reactions have, however, been found to depend on the mechanism [14]. It has been shown in Fig. 50 that in an S_N1 reaction the C-X bond is more readily heterolyzed the greater the solvent acceptor number.

The donor number is less important for the polarization of the C-X bond. On the other hand, the rates of S_N2-reactions are increased as the solvent acceptor number is decreased, because in this way, the incoming ligand Y^- is less solvated and more readily coordinated [15].

3. Formation of Reactive Cations

In a solvent of weak donor properties, ionization will be favoured by high acceptor properties, by which anions are stabilized by solvation. In such solvents of low donor number and high acceptor number, reactive cations may be produced. The strong Lewis acid anitimony (V)-fluoride has been used by *Emeléus* et al [16,17] for the formation of unusual cations, such as the BrF_2^+ and the IF_4^+ cation:

$$BrF_3 + SbF_5 \rightleftharpoons BrF_2^+ + SbF_6^-$$
$$IF_5 + SbF_5 \rightleftharpoons IF_4^+ + SbF_6^-$$

Likewise, the trimethylcarbenium ion (t-butyl cation) has been obtained by *Olah* et al [18] by dissolving t-butylfluoride in antimony(V)-fluoride, which served both as a strong Lewis acid and as a solvent. It has later been shown that the reaction proceeds according to the equation

$$(CH_3)_3CF + 2 SbF_5 \rightleftharpoons [(CH_3)_3C]^+ [Sb_2F_{11}]^-$$

Numerous other carbenium ions have been obtained [19] by using so-called "magic acids", such as $FSO_3H - SbF_5$, which have stronger acceptor properties than concentrated sulphuric acid [20]. Many other carbenium cations (with three-coordinated carbon atoms), as well as carbonium ions (with tetra- and pentacoordinated carbon atoms) have been prepared, characterized and their role in the course of electrophilic reactions carefully investigated [21]. The further development led to the formation of "superelectrophiles" [22].

The rate of solvolysis of t-butylchloride is increased as the solvent acceptor number is increased, apparently due to favourization of the formation of the carbenium cation [23].

4. Ion Association

Solvent-Separated Ion Pairs

About 60 years ago, the success of the elementary electrostatic theory by *Debye* and *Hückel* [24], extended by *Onsager*[25], to the calculation of activity coefficients and transport properties of ions led to the view that interactions between ions can be considered as due to electrostatic interactions. *Bjerrum* [26] and later *Fuoss* [27] related ion-pair dissociation constants to the dielectric constants of the solvents[1].

---

[1] *Bjerrum* treated the problem of ion-pair association from purely electrostatic considerations and defined a distance between oppositely charged ions arbitrarily, so that the work needed to separate them is four times the mean kinetic energy per degree of freedom [28].

About 30 years ago, the results of many investigations showed, however, that the results obtained on non-aqueous solutions cannot be adequately represented by elementary electrostatic models [4,29,30]. The successful application of the donor acceptor approach to ionization phenomena in solutions, requests also its application to ion-pair equilibria. Ion-pair formation is considered as different from the formation of an unionised species, because the solvated ions retain essentially their inner solvation shells. The result of this association is consequently called a "solvent-separated ion-pair".

$$A_{sv}^{+} + B_{sv}^{-} \rightleftharpoons [A_{sv}^{+}.B_{sv}^{-}]^{0}$$

The spectral properties of ionized and ion-paired species are similar because both of them are solvated, but they may be distinguished by their different conductance behaviour. By combination of spectroscopic and conductometric methods an enormous number of stability constants has been obtained [31-33].

The interpretation of many data obtained in aqueous solutions proved difficult in terms of the electrostatic theory. Their application to solutions in organic solvents was found even more difficult and the description in terms of solvent separated ion-pairs hardly possible, because of their poor solvating properties.

Contact Ion Pairs

Winstein [34,35] introduced to organic chemistry the concept of contact ion pairs or intimate ion pairs. These were originally postulated in order to explain the stereochemical course of solvolysis and electrophilic substitution reactions. The results of UV-spectroscopy and of electron spin resonance-spectroscopy were also used in support of this concept [36-38].

As a distinction is made between a *contact ion pair* and an *unionized species*, a kind of "memory" ought to exist to the constituents from which a bond is formed. A hydrogen fluoride molecule prepared by a neutralization reaction might be considered as a contact ion pair as distinguished from the covalent substrate produced by the interactions of the elements [4].

This distinction has been advanced for metal-fluorenyl compounds [37], but it has been shown that within the limits of experimental accuracy such a distinction cannot be maintained. The extent of ionization depends on the acceptor properties of the metal ion and on the solvent donor number, rather than on the dielectric constant of the medium and the IR-frequences are nearly independent of the anion in a solvent of high donor number such as dimethyl sulfoxide or pyridine [44,45].

Nearly constant frequencies, independent of the anion, are found for lithium salts in a mixture of benzene and dimethylsulfoxide, ranging in dielectric constant from 7 to 46. This shows that the preferential coordination of lithium ions by DMSO molecules is independent of the dielectric constant of the medium, an increase of which should favour the separation of contact ion pairs.

Table 32. Amount of solvent separated ion pairs (and free ions) of alkali metal salts of 9-
fluorene (Fl) in weak donor solvents at 25°C.

Solvent	DN	DEC	Li^+Fl^-	Na^+Fl^-
Dioxane	15	2.2	0	0
Toluene	-	2.4	0	-
2-MeTHF	20	6.3	25	0
THF	20	7.6	75	5
glyme	20	7.2	100	95
DMSO	30	45.0	100	100
PY	33	12.3	100	100

According to the donor-acceptor approach the formation of a contact ion pair has to be considered as a substitution reaction in which solvent molecules are eliminated from cation and anion [30].

A boundary case of contact ion pairs may be considered, for example for tetraalkylammonium tetraphenylborate, because cation and anion are hardly solvated. For tetraalkylammonium halides solvation of the anions must be taken into consideration. Their association constants in alcohols decrease in the order $I^- > Br^- > Cl^-$ in contrast to the expectations from electrostatic theory, but in agreement with the interpretation of hydrogen bonded interactions. These increase as the donor properties of the halide ion increase.

In the alcohols their association constants become increasingly differentiated with increasing cation size. In methanol, the association constants for tetramethylammonium bromide and iodide differ by a factor of 1,3, but those of tetrabutylammonium bromide and iodide by a factor of 5.

The reverse trend is observed in the gas phase and in aprotic solvents. In acetonitrile tetramethylammonium bromide is more strongly associated than the iodide, but the tetrapropylammonium salts have almost the same association constants. This has been explained by the weaker solvation of anions in aprotic solvents owing to lack of hydrogen bonding [30].

In acetonitrile the tetramethylammonium halides show increasing association constants in the order $I^- < Br^- < Cl^-$, whereas the tetrabutylammonium salts show nearly the same association constants. This levelling effect becomes more pronounced as the solvent acceptor number is increased [30].

This may be illustrated by the influence of the cation as a supporting electrolyte on the polarographic half wave potentials of the redox system hexacyano-ferrate(III)/(II) in solvents of different acceptor number [41,42]. The redox potentials are shifted to more positive values as the solvent acceptor number is increased and as the size of the cation of the supporting electrolyte is increased, i.e. in going from Et_4N^+ to Bu_4N^+, but the differences become smaller as the solvent acceptor number is decreased.

Table 33. Half wave potentials $E_{1/2}$ for the redox couple $[Fe(CN)_6]^{3-}/[Fe(CN)_6]^{4-}$ in 0,1 molar solutions of tetraethylammonium perchlorate (TEAP) and tetrabutylammonium perchlorate (TBAP) as supporting electrolytes in solvents of different acceptor properties at 25°C [42].

Solvent	$E_{1/2}$ [V] TEAP	$E_{1/2}$ [V] TBAP	$\Delta E_{1/2}$ [V]	AN	DEC	μ[Debye]
DMF	- 0.31	- 0.61	- 0.30	16	37	3.86
An	- 0.28	- 0.42	- 0.14	19	36	3.96
DMSO	- 0.27	- 0.39	- 0.12	19	47	3.90
PDC	- 0.17	- 0.20	- 0.03	18	65	4.98
NM	- 0.07	- 0.14	- 0.07	21	37	3.57
EtOH	+ 0.30	+ 0.24	- 0.06	37	24	1.70
MeOH	+ 0.47	+ 0.38	- 0.09	41	33	1.70

This is because the extent of association between the tetraalkylammonium ions and the complex anions decreases as the outer-sphere solvation of the highly basic anions is increased. This effect is more pronounced for the (strongly associated) Et_4N^+ salts than for the corresponding Bu_4N^+ and with increasing solvent acceptor strength solvation is favoured in competition with the cation coordination.

Other Hydrogen Bonded Features

Ion pairing by outer-sphere coordination may involve hydrogen bonding. In the process

$$[Co(en)_2Cl_2]^+_{sv} + Cl^-_{sv} \rightleftharpoons [Co(en)_2Cl_2].Cl$$

solvent molecules co-ordinated in the outer-sphere of the complex ion are replaced more readily by chloride ions as the solvent donor number is decreased [39]. Ion pairing through hydrogen bonding is established as the solvent donor properties are less developed than those of the competing chloride ions, which are increased by decrease in solvent acceptor number. Hence the formation constant of the ion pair is greater as *both* the solvent donor number *and* the solvent acceptor number is decreased [43,44].

The outer-sphere effect is in agreement with the results of polarographic measurements [46].

An example of ion pairing by means of hydrogen bonding at the cationic species is provided by the behaviour of trialkylammonium salts in different solvents. Tetrabutylammonium picrate is non-associated in solution of nitrobenzene [47], where the tributylammonium salt has an association constant of 526. This dissociation involves replacement of the hydrogen bonded anions by donor solvent molecules D:

$$Bu_3NH^+ \leftarrow Pic^- + n D \rightleftharpoons [Bu_3NH \leftarrow D]^+ + [Cl.D_{n-1}]^-$$

Table 34. Formation constants for the 1:1 outer-sphere complex from $[Co(en)_2Cl_2]^+$ and Cl^- [45].

Solvent	K_{form}	DN	AN	DEC
Methanol	150	34	36	33
DMSO	400	30	19	47
DMF	8000	27	16	37
DMA	20000	28	14	38
TMS	42000	15	19	43

Accordingly, the dissociation constant in acetonitrile (DN = 14, DEC = 36) is 40 times greater than in nitrobenzene (DN = 4, DEC = 36) [48]. Likewise, in trialkylammonium chloride replacement of a hydrogen bonded chloride ion by solvent molecules is favoured by increase in solvent donor number (increased stability of hydrogen bonding with solvent molecules) and by decrease in solvent acceptor number (decrease in solvation and greater donor property of the chloride ion for hydrogen bond formation)[49].

In solvents of low donor number and low acceptor number, evidence is available for the presence of so-called triple ions, even in media of appreciable dielectric constant, such as nitrobenzene (DEC = 36), which may be formulated by hydrogen bonding: $[R_3NH^+ \leftarrow Cl^- \rightarrow {}^+HNR_3]^+$.

5. Less Common Aspects of Non-Aqueous Solvents

Non-Aqueous Micelles

Little attention has been drawn to the existence of micelles in non-aqueous solutions, probably because non-aqueous solvents are much more reluctant to form them.

Whereas solvents with a single hydrogen bonding centre, such as methanol, ethanol or DMF do not support micelle formation, solvents with two or more centres available for hydrogen bonding may give rise to micelle formation [49]. Examples of such solvents are ethyleneglycol, 2-aminoethanol, formic acid, 1,2-propandiol.

However, for a given surfactant, micelle formation is less favoured and less differentiated than in water, as is seen, for example, in the case of dodecylpyridinium bromide (see the comparison of the critical micelle concentrations in Table 35).

Table 35. Critical micelle concentrations (CMC) at 27°C in ethylene glycol and in water [49].

Surfactant	CMC in ethylene glycol [mol.l^{-1}]	CMC in water [mol.l^{-1}]
Dodecylpyridinium bromide	0.55	$1.22 . 10^{-2}$
Tetradecyltrimethylammonium bromide	0.25	$3.84 . 10^{-3}$
Hexadecylpyridinium chloride	0.23	$9.20 . 10^{-4}$

Another group of micelle forming solvents is provided by typical hydrocarbon solvents, such as benzene, cyclohexane or heptane. In these solvents the structures of the micelles are described as "reversed" to those in water, because the "polar" head groups are accumulated in the centre of the micelle and the "non-polar" side chains directed towards the "non-polar" solvent. This is again in accordance with the simile rule. The aggregation numbers of these inverted micelles tend to be much lower than those of their aqueous counterparts and they are also less polydisperse.

Hardly Removable Solutes

Although several statements have been made that extremely small traces of solutes may profoundly influence the properties of a liquid, their influence has not been studied in detail. Just as water cannot be freed from dissolved gases, their presence seems also unavoidable in any other liquid. The gas solubilities in non-aqueous solvents are usually greater than in water (Table 16), but *the last traces of gases have never been removed from a liquid.*

Another solute that seems to be unremovable from non-aqueous solvents is water. Many carefully "dehydrated" solvents, such as acetone or nitromethane contain 10^{-3} mol water per litre! Acetonitrile has been obtained with a water content of 10^{-4} mol/l [50] and the presence of water in most other solvents can usually be estimated by means of *Karl Fischer* titrations.

It has been suggested that at low water content the strong interactions between the water molecules are replaced by solute-water interactions with decrease of the rotational correlation times of water molecules [51-53].

References

1. J. J. Lagowski (ed.) *The Chemistry of Non-Aqueous Solvents* (Academic Press, New York, London, 1966).
2. V. Gutmann, *Electrochim. Acta* **21** (1976) 661.
3. V. Gutmann and U. Mayer, *Monatsh. Chem.* **100** (1969) 2048.
4. V. Gutmann, *Angew. Chem.* **82** (1970) 858, Int. ed. **9** (1970) 843 - *Coord. Chem. Revs.* **18** (1976) 225.
5. H. F. Ebel and R. Schneider, *Angew. Chem.* **77** (1965) 914, Int. ed. **4** (1965) 878.
6. E. Weiss and E. C. A. Lucken, *J. Organometal. Chem.* **2** (1964) 197.
7. H. L. Lewis and T. L. Brown, *J. Amer. Chem. Soc.* **92** (1969) 4664.
8. G. Normant, *Angew. Chem.* **79** (1967) 1029.
9. M. Schlosser, *Struktur und Reaktivität polarer Organometalle* (Springer, Berlin, New York, 1973).
10. W. Weiss and G. Sauermann, *J. Organometal. Chem.* **21** (1970) 1.
11. W. I. O'Sullivan, F. W. Swanner, W. J. Humphlett and C. R. Hauser, *J. Org. Chem.* **26** (1961) 2306.
12. D. J. Cram, *Fundamentals of Carbanion Chemistry* (Academic Press, New York, London, 1965).
13. U. Mayer, V. Gutmann and L. Lodzinska, *Monatsh. Chem.* **104** (1973) 1045.

14. V. Gutmann, *Chem. Techn.* (1977) 255.
15. A. J. Parker, U. Mayer, R. Schmid and V. Gutmann, *J. Org. Chem.* **43** (1978) 1843.
16. A. A. Woolf and H. J. Emeléus, *J. Chem. Soc.* (1949) 2865.
17. H. J. Emeléus and A. G. Sharpe, *J. Chem. Soc.* (1949) 2206.
18. G. A. Olah, W. S. Tolgyesi, S. J. Kuhn, M. E. Moffat, I. J. Bastien and E. B. Becker, *J. Amer. Chem. Soc.* **85** (1963) 1328.
19. G. A. Olah and A. M. White, *J. Amer. Chem. Soc.* **91** (1969) 5801.
20. R. J. Gillespie, *Accounts Chem. Res.* **1** (1968) 202.
21. G. A. Olah, *Angew. Chem.* **85** (1973) 183.
22. G. A. Olah, *Angew. Chem.* **105** (1993) 805.
23. U. Mayer, V. Gutmann and W. Gerger, *Monatsh. Chem.* **106** (1975) 1235.
24. P. Debye and E. Hückel, *Physikal. Zeitschrift* **24** (1923) 185, 305.
25. L. Onsager, *Physikal. Zeitschrift* **27** (1926) 388.
26. N. Bjerrum, *Kgl. Danske Videnske math.-fys. Medd.* **9** (1926) 7.
27. R. M. Fuoss, *J. Amer. Chem. Soc.* **80** (1958) 5059.
28. M. T. Beck, *Chemistry of Complex Equilibria* (Van Nostrand Reinhold Co, London, 1970).
29. U. Mayer and V. Gutmann, *Structure and Bonding* **12** (1972) 113.
30. U. Mayer, *Coord. Chem. Revs.* **21** (1976) 159.
31. J. Bjerrum, G. Schwarzenbach and L. G. Sillén, *Stability Constants,* 2 vol. (Chemical Society, London, 1957).
32. R. W. Gurney, *Ionic Processes in Solution,* (McGraw Hill Co., New York, London, 1953).
33. C. W. Davies, *Ion Association,* (Butterworth, London, 1962).
34. S. Winstein, E. Clippiinger, A. H. Fainberg and G. C. Robinson, *J. Amer. Chem. Soc.* **76** (1954) 2597.
35. S. Winstein and G. C. Robinson, *J. Amer. Chem. Soc.* **80** (1958) 169.
36. M. Swarc, *Ions and Ion Pairs in Organic Reactions,* (Wiley, New York, 1972).
37. T. E. Hogen-Esch and J. Smid, *J. Amer. Chem. Soc.* **88** (1966) 307, 318.
38. P. Chang, R. V. Slates and M. Swarc, *J. Phys. Chem.* **70** (1969) 3180.
39. B. W. Maxey and A. I. Popov, *J. Amer. Chem. Soc.* **89** (1967) 2230.
40. B. W. Maxey and A. I. Popov, *J. Amer. Chem. Soc.* **91** (1969) 20.
41. V. Gutmann, G. Gritzner and K. Danksagmüller, *Inorg. Chim. Acta* **17** (1976) 81.
42. G. Gritzner, K. Danksagmüller and V. Gutmann, *J. Electroanal. Chem.* **72** (1976) 177.
43. V. Gutmann, *Chimia* **31** (1977) 1.
44. G. Gonzalez, U. Mayer and V. Gutmann, *Inorg. Nucl. Chem. Letters* **15** (1979) 155.
45. W. R. Fitzgerald, A. J. Parker and D. W. Watts, *J. Amer. Chem. Soc.* **90** (1968) 5744.
46. U. Mayer, A. Kotocova, V. Gutmann and W. Gerger, *J. Electroanal. Chem.*

100 (1979) 875.

47. C. R. Witschonke and C. A. Kraus, *J. Amer. Chem. Soc.* **69** (1947) 2472.
48. P. L. Huyskens, *Private Communication.*
49. G. C. Kresheck, in *Water, a Comprehensive Treatise,* ed. F. Franks, Vol. 4 p. 140 ff (Plenum Press, New York, 1975).
50. V. Gutmann and A. Steininger, *Allg. Prakt. Chem.* **18** (1967) 282.
51. W. Zeidler, in *Water, a Comprehensive Treatise,* ed. F. Franks, Vol. 2, p. 584 (Plenum Press, New York, 1973).
52. E. Greinacher, W. Lüttke and R. Mecke, *Ber. Bunsenges.* **59** (1955) 23.
53. J. R. Holmes, D. Kivelson and W. C. Drinkard, *J. Amer. Chem. Soc.* **84** (1962) 4677.

CHAPTER 16

QUALITATIVE ASPECTS OF THE MOLECULAR CONCEPT

1. Impact and Limitations of the Present Molecular Concept

In Chapter 3 it has been emphasized that the molecular concept has been successful both on the macroscopic and on the submicroscopic (molecular) level. On the former it accounts for the quantitative aspects encountered in the course of chemical changes. Its enormous success on the molecular level is due to the great number of *quantitative relationships* between various molecular properties and their changes within restricted dimensions. Such relationships are found in all areas of chemistry, notably in solution chemistry and in organic chemistry as well as more recently in molecular biology and supramolecular chemistry.

On the other hand, these relationships do not provide an understanding of the *qualitative changes* that occur in the course of chemical interactions. It has been pointed out in Chapter 3 that the "chemical equation" is bound to be restricted to equalities in numbers [1], because it would be impossible to find equalities of qualities. The same applies to the application of the molecular concept on the molecular level as it does not explain the qualitative features.

Because modern chemistry is mainly interested in establishing quantitative relationships, its successful application to them is greatly appreciated, whereas its *failure to account for qualities* and their changes seems to provide no serious problem to modern science.

The situation has been characterized by *Primas* [2] as follows: "The molecular view has triumphed in physics, chemistry and biology with immense practical results. In the main, chemistry has fulfilled its molecular program. Our main thesis is, that the richness of chemical phenomena renders it impossible to discuss them exhaustively from a single point of view".

The authors of this book feel that the exhaustive characterization of chemical phenomena is not possible from the single point of view of quantitative considerations. The need for the inclusion of qualitative considerations has been well demonstrated for example by the success of the joint efforts of qualitative and quantitative considerations in the course of the establishment of the periodic table of the elements about 125 years ago (see p. 20).

We propose therefore to raise the question, as to whether the molecular concept is suitable for the elucidation of the qualitative character of the phenomena.

2. Quality and Quantity

All philosophical systems agree that there is an equality with regard to substance and this equality is called *Identitas*. This is at the basis of all philosophical considerations and is aimed at a complete adaptation of thought to the object that is being perceived and studied. The aim of identitas is a complete conformity of intelligence and reality. Because *substance is sensually not perceptible* (although accessible through quality), it is not considered in modern scientific investigations.

On a *quantitative* level we find *Aequalitas*, the equality in numbers. Natural science is based on this, with its quantitative measurements and its mathematical description. Aequalitas, as used to express equalities in numbers, is frequently helped in science by statistics (see p. 42).

Finally there is the *Similitudo*, the similarity of form, in spite of and because of individual differences in *qualities*. This is the fundament of art and of qualitative science. With similitudo one gains knowledge about similarity of form and about quality. Quality is *"form in motion"* . There can never be aequalitas of quality, since this would demand motionless (timeless), strictly defined points.

It has been pointed out on p. 1 that quality is the expression of the object as it is perceived by our senses and that quality is the requirement for quantities to be measured.

In contrast to the human senses, a physical measuring instrument is extremely limited with regard to its responsiveness to varied qualities. An instrument can usually respond to one kind of quality only. A barometer, for instance, can be used to measure the gas pressure, but not the temperature, the chemical structure or the conductivity. A physical measuring instrument can only partially react to changes in a system, it can illustrate one qualitative aspect. Actually it is for the very reason of its limitations that the physical measuring instrument supplies, in the case to its responding to a quality, precise results with regard to the quantitative aspects of the quality.

The enormous "advantages" of quantitative data are not only measurability and precision, but also divisibility, separability from the object and - last but not least - it allows disposability about them by man. For these reasons, scientific methodology has been developed to deal with measurable and divisible quantities and not with unmeasurable and indivisible qualities.

In fulfilment of *Galilei's* advice, mentioned on p.3, to "measure whatever can be measured and render measurable what cannot yet be measured", *modern science has attributed the primacy to quantity rather than to quality.*

Both, quality and quantity can be found on each object and they cannot be completely separated from each other. As our senses are moved by the qualities[1] , these provide the starting point and the *primary source for the acquisition of knowledge of all natural things and events:* an unmeasurable quality is the prerequisite for the quantity to be measured.

We do not suggest the exclusion of quantitative considerations, but we feel obliged to suggest that one should discontinue the overemphasis on quantity in

---

[1] Because motions cannot be described by mathematics, the mathematical formulation of qualities is not possible (see p.3).

modern science and in this way pave the way for a "renaissance" of qualitative science.

In order to realize similarities of quality, the method required is based on comparison and judgement of similarities. Similitudo is a relationship between things that is based on the Unitas Qualitativa, i.e. *the unity of quality* (in German: "Einheit der Beschaffenheit"). This does not demand unity of the parts. It is not always realized that all knowledge of man, his realizations and perceptions are based on similarity considerations.

Similarity considerations of qualities lead to analogous terms, which are the last concepts of philosophy. The role of analogy considerations in the development of chemistry has been pointed out in Chapter 1.

One can compare things that have more in common than a mere name. This is why nominalists, who accept nothing but names, can never arrive at qualities and at a substance [3]. They consider quality only as a name for a relationship between subject and the quantitative aspects on the non-subjective side.

3. Starting Points for Qualitative Investigations

It has been emphasized on p.19 that the properties of the parts which are characteristic for them within the more complex unit are irretrievably lost by its analytical dismemberment [4]. Each attempt to "reconstruct" the more complex unit from the properties of more or less isolated parts relies on the imaginations of the investigator and leads to the (premature) formulation of a model or of a theory, which for this reason always contains elements of speculation. Attempts must be made to check the validity of any imagination by experiments which are made under artificially chosen conditions. The experimental results reveal certain inconsistencies of the assumptions and hence the original imaginations have to be modified or to be replaced by others and these are bound to show other shortcomings. As this procedure is continued, it leads to development "ad infinitum".

We suggest therefore, that the correct procedure, namely to start from the properties of the complex structures under various conditions, to learn as much as possible about them and to proceed from this knowledge to the simple structures in order to draw conclusions about their "abilities" from their properties within the complex relationships, because these abilities cannot be found outside the complex conditions. In other words: *the unfolded explains the enfolded and not the other way round.*

Although this approach is difficult, it cannot be completely avoided in studies of chemical changes, where each selected entity must be studied under certain environmental conditions. For example, the properties of an acid may be studied by interactions with bases and the reducing properties of a reducing agent in the presence of oxidizing agents. Each investigation is normally carried out with respect to a certain question and simplified experimental conditions are "appropriately" chosen in order to avoid complexity as much as possible.

Even under such restricted circumstances, some of the "abilities" of the entities are found, although other abilities, which are developing only under more complex conditions, cannot be disclosed in this way. This may be illustrated by the following examples:

(i) The fate of the polarity scale of the elements, proposed by *Berzelius* (and mentioned on p.20) is a consequence of the emphasis on the properties of the elements in states of "splendid isolation". Under these conditions the atoms are rightly considered as non-polar and the development of polarities under other conditions cannot be found, unless investigations are carried out under such conditions. As long as the development of polarities had not been explicitly taken into consideration, the atoms were considered non-polar and their characterization by means of polarities was rejected.

(ii) In the attempt to relate molecular structures to chemical reactivities, wrong conclusions have often been drawn; for example for the different Lewis acidities of the boron halides. By comparing the structural features and the electron densities of boron(III)-fluoride and boron(III)-bromide, stronger Lewis acidity would be expected for boron(III)-fluoride, because of its smaller electron density at the unoccupied 2p-orbital and the stronger polarity of its bonds. Experiment revealed, however, that the bromide is the stronger Lewis acid, and this strongly suggests that the abilities for charge transfer and polarizabilities cannot be accounted for by considering the properties of the isolated molecules.

(iii) When argon and copper are studied in the gas phase, a distinction between rare gas atoms and metal atoms cannot be made, because each of the gases is monoatomic and neither of them shows any of the properties as that are characteristic of a metal. In order to make this distinction, the gases must be condensed and the properties of each of them studied in the solid state. It is not enough to determine their crystal structures, because both of them crystallise in the same cubic face-centred type of structure (which in most textbooks is described as a metallic lattice). After consultation of all of their properties in the solid state, such as conductivity, ductility etc. the conclusion may be drawn that copper atoms in the gas phase must have abilities to reach the metallic state and hence to consider them as metal atoms. Because such properties are not found for solid argon, it must be concluded that the argon atoms are lacking these abilities and hence they are considered as non-metal atoms.

Likewise, the different abilities of iron and ruthenium atoms (with their similarities in electron configurations) are revealed by comparing the properties in the metallic states. This leads to the conclusion that iron atoms have abilities to form "ferromagnetics" in the solid state which ruthenium atoms are lacking.

(iv) A final example may be provided by the differentiation of the three sets of p-orbitals of given principal quantum number. No experimental evidence is available from spectroscopic data for this differentiation, because all of these p-electrons are found in the same energy states, and therefore cannot be differentiated on spectroscopic grounds alone. The only experimental justification for the construction of three different p-orbitals is provided by the splitting of the spectral lines in a magnetic field. The description of this differentiation is maintained for the fictitious isolated atom, i.e. in the total absence of a field and of interacting forces. This differentiation is justified because their actual behaviour in the magnetic field shows these abilities, but this conclusion would have been impossible on ground of studies in the absence of a magnetic field.

These examples illustrate the necessity

(i) to start from studies under complex conditions,

(ii) to proceed to the "potentialities" of the parts under less complex conditions and

(iii) to reach conclusions about the abilities of the parts which are based on observations and not on our imagination[2].

In order to learn as much as possible about the "abilities" of the parts, we have to start from the observations under the most complex situations possible and not from artificially simplified experimental conditions (p.171). For example, in order to study liquid water, we are accustomed to start from results obtained from studies of single water molecules, to make assumptions of their interacting forces and to invent models, which are then tested by performing experiments under "adequately chosen", i.e. reduced conditions (Chapter 5).

Instead of this approach we propose starting from the properties and functionalities of liquid water within the highest and most complex organizational form on earth, namely the human body, where water appears to provide for the unity of the human being, for the storage and transfer of matter, energy and information in highly specific ways (Chapter 19).

4. Dynamically Ordered Relations

The principle that all things are in constant motion was first expressed by the ancient Greek philosopher *Heracleitus* (540 - 480 B.C). He considered that the resulting dynamic equilibrium maintains an orderly balance in the world and expressed in this way the view, later formulated by *Aristotle*, that *order is regularity between before and after*. The persistence of unity by means of changes, he illustrated by analogy to a river, when he said: "Upon those who step into the same river, different and ever different waters flow down".

The *Aristotelian principle of motion* is, like any other principle in nature, outside of matter and space, it is related to the being of all things, but it cannot be described. It can be described only according to abstractions as a *conditio sine qua non, that all in nature is subject to dynamically ordered relationships*.

---

[2] Supramolecular Chemistry as recently advanced [5] is defined as chemistry beyond the molecule and concerned with organized entities of higher complexities that result from the association of two or more chemical species held together by intermolecular forces. *These are based on our imaginations*. In these ways attempts are made in order to design molecules with special properties and to invent novel species and processes. The use of hardly described terms such as "purposeful bonding", "molecular recognition", "selection of information" (see also the remarks by *Weiss* in Chapter 19) remind the authors of this book to the following statement by *Weizsäcker* [6]: "One of the methodical foundations of science lies in the avoidance of certain fundamental questions. It is characteristic of physics, as practised nowadays, not to really ask what matter is, for biology, not to really ask what life is, and for psychology not to really ask what soul is. Instead these words paraphrase vaguely a field in which research is intended. This fact is a methodical prerequisite for the success of science... On the other hand, one cannot be deceived that this methodical approach, if the researcher is unaware of its doubtfulness has an internecine quality about it".

From the principle of motion and from the condition of dynamically ordered relationships, it follows that each of the observed things acts as a unity in its meaningful response to stimuli from outside. Such activities, or rather reactivities have been pointed out on p.20 as implied in the most fundamental laws of physics. These imply that *each object must be able to respond in highly specific ways towards changes in environment, whereby its main characteristics and functionalities are not lost.*

Such abilities and activities of molecules have been expressed by the functional approach described briefly on p.26 ff. and by a kind of "intelligent behaviour" of biologically active molecules on p.36 and these imply system characteristics (see p. 229 ff).

5. System Considerations

The word system is of Greek origin and expresses an activity of "putting together different parts to a whole", which acts as a unity. This requires that the parts are not arbitrarily arranged but, according to the requirement of the whole, form a set of specifically interacting parts of a material or of an immaterial thing.

With regard to material systems, the arrangement of the parts may be provided in nature or assembled by man. In the first case the system is called a *natural system* and in the second case an *artificial system*.

The whole universe is considered as one natural system and this is described by quantum mechanics by means of a fluctuating charge density pattern.

In the universe different parts can be recognized as they are observable as discontinuities within the continuum (see Chapter 2).

Within the universe the heliocentric system has been well established with its various objects, namely the sun, the planets, the moons and the planetoids. These are recognized by means of their boundaries, but there are no recognizable boundaries for the heliocentric system as such. In order to find out which of the said objects belongs to the system, the dynamically ordered relationships must be taken into consideration. The motions of the planets are found under the dominating influence of the sun, which has a mass about 1000 times greater than that of the biggest planet, the Jupiter. Each of the moons is moving under the influence of one of the planets and hence is part of the whole system.

Each of these objects, the sun, the planets, the moons and the planetoids may be considered as one of the subsystems. Apart from the earth, they can be recognized by man by means of their boundaries, but it is impossible to define the boundaries of the earth. The atmosphere certainly belongs to the earth, but there are no observable borderlines. Again, the boundaries may be inferred from the dynamic relationships such as those between earth and moon and the mutual influences of their gravitational fields. Even these borderlines are not fixed, but rather changing as the relative positions of earth and moon are changing. It is just as difficult to learn about the internal regions of the earth, because no perceptions can be made about them. Indirect information can be obtained from their dynamic actions such as earthquakes and volcanic eruptions.

Based on the perception of boundaries, many different objects can be recognized as *subsystems on earth,* such as gases, liquids and solid materials, among them various living beings. Each of the objects or subjects appears separated and yet interconnected within its environment by means of its respective phase boundary (see p. 80).

Again, the borderlines of the objects are not clear-cut. If we look at the clouds from the earth they appear contrasted to the non-differentiated background of the sky. If we pass such boundaries, for example in an aircraft, we will find a more or less broad area in which a beginning or an end cannot be clearly recognized. It has been pointed out on p.80 that the situation is similar for the phase boundary of a solid or a liquid phase. The latter is integrated in its surroundings and yet is representing a unit of its own.

Within these subsystems even smaller subsystems may be recognized - for example the cells of an organism. Molecules and atoms may be recognized with the help of certain techniques and it has been pointed out on p.16 ff. that they remain always integrated within the continuous relationships of one of the superordinated systems of a gas, a liquid or a solid material.

In order to obtain an understanding of the quality of a liquid, we will try to investigate the regularities and the interrelationships of the molecules within the superordinated system of the liquid under consideration.

For the time being, the following similarities between the heliocentric system and a liquid system may be mentioned, namely

(i) the boundaries of the systems are not clear-cut and follow from the consideration of the dynamic interrelationships,

(ii) characteristic superordinations and subordinations exist between the different parts.

6. Order and Finality

Without order, no regularities would be observable on which knowledge is based. *Order* has been defined by *Aristotle* as *"regularity between before and after"* and not, as maintained by the nominalists, a construction obtained by means of abstractions. According to the nominalist view, our concepts, like that of order, are just names and bare no relationship to the things as they are, i.e. all that can be recognized and perceived never corresponds to a thing as it is.

This view allows and encourages the definition of concepts by means of the imagination and fantasy of the investigator. We have become accustomed - although frequently without being aware of it - to use names obtained in these ways with the result that we cannot reach the real things, but rather our own model assumptions, to which we get accustomed.

Following the nominalist view, in modern physics the "state of perfect order" is attributed to the result of human imagination, namely to the ideal crystal at absolute zero temperature (see p. 41). In this way order is attributed to the fiction of an absolutely rigid crystal without any dynamic features, to something incapable of existence in nature, to the "mere name of the ideal crystal".

Other nominalistic definitions in modern science are, inter alia, the model of the ideal gas (which would be devoid of all static aspects of order), the vacuum as considered as completely empty (which would be outside the continuous relationships and therefore unobservable and unmeasurable, see p.11), the closed system (which again would be outside the continuous relationships, see p. 4), the reversible process (which would involve the reversibility of time), the concept of an isolated atom (which would be outside the natural relations and hence unobservable), the orbital concept (which has been introduced for calculations in an abstract manner) and all of the assumed intermolecular forces such as *van der Waal's* forces (p. 41).

In this way, the question of the purpose of order has been excluded according to the demand by *Francis Bacon* (1561-1626), who stated: "The study of natural processes under the aspect of their telos is sterile like a consecrated virgin, it gives birth to nothing". He considers the teleological question as sterile and not as wrong and judges a virgin only from the point of view of the possibility of her giving birth.

Johannes Kepler emphasized in 1611 [7], that a study of bodies and matter entails a study of their purpose, to which he referred in a study on the six-cornered snow flake to honeycombs in the following way: "It is certain that bees shape their own combs and themselves build up their many-storied blocks from the foundations. The bee, therefore, by nature has this instinct as its property to build in this shape rather than others. This original pattern has been imprinted on it by the Creator. The matter of wax or of the bee's body can have nothing to do with it; the process of growth nothing either. This observation at once raises the further question of the purpose, not which the bee itself pursues in its business, but which God himself, the bee's creator had in mind when he prescribed to it these canons of its architecture. Now here again at long last a consideration of bodies and matter enters into the determination of purpose".

Bohm [8] has pointed out the central problem in physics to redefine the concept of order, although without touching the teleological question: "Original and creative insight within the whole field of measure is the action of the immeasurable. For, when such insight occurs, the source cannot be within the ideals already contained in the field of measure, but rather has to be in the immeasurable, which contains the essential formative cause of all that happens in the field of measure. The measurable and the immeasurable are then in harmony, and indeed one sees that they are different ways of considering the one and undivided whole ... We proposed that a new notion of order is involved here which we called the implicate order (from the Latin root meaning "to enfold" or to fold inward). In terms of the implicate order one may say that everything is enfolded into everything. This contrasts with the explicate order now dominating in physics in which things are unfolded in the sense that each thing lies only in its own particular region of space and time and outside the regions belonging to other things ... When one works in terms of implicate order, one begins with the undivided wholeness of the universe."

In classical physics, the actually occurring relations are expressed within the framework of mechanistic order, the *causal-deterministic view*. This allows physically relevant conclusions by considering relationships between initial and resulting states. However, each initial state results from a former initial state and each resulting state is at the same time the initial state for the next causal step to follow. This shows that causality requires the teleological horizon. Each of theses states must be seen within the whole context. The finality is - according to *Aristotle* - the supreme cause of order.

Spaemann and *Löw* [9] make the following comments: "The reduction of our knowledge of nature to its causal explanation is the result of our wish to rule nature. In order to meet this artificial purpose, all attempts at teleological considerations of reality have been discarded. Modern science uses only dependent variables in regarding an event as due to a previous one in a causal sense, and hence it expresses only passivity".

According to *Aristotle*, finality in the natural things is inherent in them. He calls it "entelechia", meaning "I carry the finality within me". This means that we have to try to find the ordered relationships in nature and not to construct them with the help of our imagination.

In order to illustrate the problem of gaining access to the purpose, we may consider an artificial system, i.e. a system constructed and operated by man [10]. A computer serves a certain purpose and so does each of the programs, the software. These have been determined and worked out by man. Neither purpose nor software can be directly observed, but adequate knowledge can be obtained from those who designed the computer and the software and the relevant directories.

With regard to the investigation of the ordered relationships in nature, the investigator finds himself in a position similar to that of an experienced operator at a computer, after this has been loaded with a software unknown to him. He has to rely on the messages which appear on the screen and he obtains some indications about the kind of software by similarity considerations with programs familiar to him.

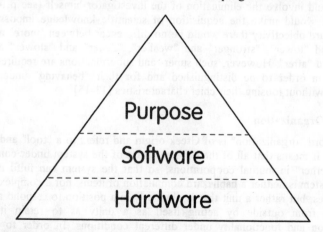

Fig. 55. Illustration of the main hierarchic levels of a computer (in fact the hierarchic levels indicated in this figure are further differentiated).

He may distinguish different kinds of software, such as word processing, data processing, business calculations, graphic program, program language, data operating system and computer games. In this way he obtains some information about the purpose of the program. Further details about its functions can be obtained by laborious trial and error methods and analogy considerations, but he may find it extremely difficult to obtain complete knowledge about a complex program.

We may be reminded of the following words of *Robert Louis Stevenson* [11]:

"Who has seen the wind?
Neither you nor I.
But when the trees bow down their heads,
The wind is passing by."

One of the difficulties in expressing such relationships is caused by the limitations of the language used in modern science. As *Weiss* [12] has pointed out, as a tool for the articulation of knowledge, scientific language must be kept in step with advances of knowledge. The language used in science has been developed to express in the first place quantitative aspects within the frameworks of causality and probability considerations. However, in order to describe qualitative changes and system behaviour, it will be necessary to introduce some new vocabulary to science, such as situation, capability, adaptability, vulnerability, flexibility, memory, control, subordination, regulation, static and dynamic aspects of order, coexistence, balanced interactions, integral configuration, molecular system organization, supermolecular system organization, connectivity, cooperativity etc. [13].

At this point attention may be drawn to the anthropomorphic elements which have just been used. However, an attempt at complete exclusion of them from science would involve the elimination of the investigator himself (see p. 48 f), and therefore it would make the acquisition of scientific knowledge impossible. In a world of pure objectivity there would be no differences between "more" and "less", "higher" and "lower", "stronger" and "weaker", "faster" and "slower" as well as "before" and "after". However, such super- and subordinations are required both for the things in order to be distinguished and for their "behaving" under different conditions without loosing their chief characteristics [13-15].

7. System Organization

The word "organization" is of Greek origin and refers to a "tool" and as a verb "to work". It means that all of the different parts of the system under consideration "work together" in mutual cooporations, so that the system can fulfil its task. A material system is neither a haphazard compilation of items, nor a complex of rigidly linked pieces, but rather a unit that is obviously in a position to respond adequately to actions from outside by acting itself as a unity as to retain its integral configuration and functionality under different conditions. In order to make this possible, *a system organization is a requirement for each real thing and not the result of our imagination* [16].

The essentials of the organization in living systems have been described by *Weiss* [12] in his book "Hierarchically Organized Systems in Theory and Practise" as follows: "As a unit is composed of subunits, the problem of the unity of a unit resolves itself into the question of what makes the component subunits cooperate in such coordinated manner as to establish and preserve the rather definite pattern of the compound unit. It may be a standard pattern of behaviour of the components that yields recurrently the same unitary result even though there is no geometric similitude among the constellations of the components from movement to movement. Such units are properly called systems ... The state of the whole must be known in order to understand the coordination of the collective behaviour of the parts ... A system could, therefore be defined as a complex unit in space and time so constituted that its component subunits, by systematic cooperation, preserve the integral configuration of structure and behaviour and tend to restore it after non-destructive disturbances."

In his book "The Science of Life", *Weiss* [14] concluded, that "the patterned structure of the dynamics of the system as a whole coordinates the activities of the constituents. In atomistic microdeterministic terms, this coordination would have to be expressed as follows: Since any movement or other change of any part of the system deforms the structure of the whole complex, the fact that the system as a whole tends to retain its integral configuration implies that every change of any part affects the interactions among the rest of the population in such a way as to yield a net countervailing resultant; and this for every single part. Couched in anthropomorphic language, this would signify that at all times every part "knows" the stations and activities of every other part and "responds" to any excursions and disturbances of the collective equilibrium as if it also "knew" just precisely how best to maintain the integrity of the whole system in concert with the other constituents."

The first property of the things as they are is their *Unitas*, which allows the meaningful response to stimuli from outside. This requires complexity and differentiation as a source for the dynamically ordered relationships, as well as for the cooperativities between all of the different parts in order to provide "harmony".

This implies that the different properties and abilities of the parts cannot be accidental, but must be intentional with respect to the constancy of the unity.

Such constancy is only possible in a hierarchically structured and organized system and this leads us to the conclusion: *"Any real system is subject to a system organization"* [16]. This is the result of the careful philosophical analysis of facts, in agreement with the principles of nature and not of our own imagination.

A principle is outside of matter and space, outside the order of motion and yet, the source of all motions, the fundamental intellectual realization of the things (see footnote on p. 50). A principle can be found only by means of induction, i.e. the method to find unity in the multiplicity.

Descriptive terms can be found only as abstractions are introduced. According to them, *hierarchy is a "conditio sine qua non"*, the irreplaceable requirement for observations of ordered relationships[3]. In such descriptive terms, hierarchy is subject to scientific investigations, which deal with the conditions and with the modalities on a material basis.

The difference between a principle and a conditio sine qua non may be illustrated by the consideration of the *Pauli*-principle. The differentiation of all electrons is an expression of the differentiation according to the principle of motion, which is the source for all ordered relationships. The precise formulation in physical terms requires the introduction of abstractions, such as quantum numbers and in this way the *Pauli*-principle refers to the conditio sine qua non.[4]

Likewise, system organization cannot be described as a principle, but it can be investigated as conditio sine qua non of hierarchically ordered relationships [13].

The problem of the acceptance of the concept of system organization has been described by *Weiss* as follows [20]: "One honestly cannot deny that hierarchical order and organization as superordinated principles *sui generis* would have gained scientific acceptance more readily, if they had not been suspected of teleological implications." This means that this principle has been rejected a priori on ideological grounds.

Different regions within a material serve different purposes and hence they differ in significance for the whole system. In principle, it would be impossible that all parts serve the same purpose, for this would require that they have exactly the same environment [13]. As each of the parts has a somewhat different environment, the parts are bound to differ in properties and hence it is impossible to find two identical parts, as expressed by the *Pauli*-principle.

In order to avoid the leading influence of our imagination on the course of the investigation (see p. 49), we do not propose to apply or to develop system theories. These are successfully used in modern technology [17] and have also been recommended for the description of biological systems [18]. System theories allow quantitative applications under certain conditions, but they do not provide appropriate access to the forms of organization of natural systems.

For this reason we do not propose to follow synergetics [19], which describes some of the complex features with the help of non-linear equations. This method is well established in modern science, but it provides artificially constructed aspects of order and leads to certain misinterpretations of hierarchy, such as "enslavery".

In attempting to study the different tasks of the different parts for the whole system, it must be borne in mind that the differences exist in their continuous relationships, and that they can be perceived only to a limited extent. Differences can be perceived only with the aid of observable discontinuities, which are required for both observation and measurement (Chapter 2). In order to learn about differences in

---

[3] A conditio sine qua non is on a material level and therefore subject to scientific investigation; it remains always within the description of the concrete things and events.

[4] Another conditio sine qua non is the entropy as the quantitative expression of the irreversibility.

the absence of observable discontinuities, we have to make use of abstractions by introducing artificial discontinuities.

Before proceeding to this step, it may again be emphasized that any natural system is subject to a hierarchic system organization [16]. This means that those who feel sceptical about hierarchy or who are even opposed to hierarchy as a concept are actually sceptical or in opposition to nature. This attitude is, however, understandable, as

(i) hierarchies in nature have not been subject to scientific investigations and

(ii) all information about hierarchies is taken from that provided by artificial systems and their faults due to their construction by man. They have been established according to certain imaginations and they are executed by man. Their faults are not due to hierarchy itself, but rather the consequences of faultily designed and not properly executed orders by man.

8. Introduction of Hierarchic Levels

In the attempt to provide a description of system organizations, abstractions have to be introduced in order to provide artificial discontinuities within the continuous relationships.

Flury [21] has emphasized that "the process of induction requires abstraction as a function of the intellect. If the intellect is able to grasp a thing, ballast must be thrown over bord, and numerous elements belonging to the world of sensation have to be left behind. Abstraction is essential to the intellect - a necessary evil - because it makes reality thinkable and expressible. This is a sign of poverty: to save the essential and to reach a conception, which is related to the things as they are, one needs to be prepared to sacrifice numerous *accidental* elements".

We have to make use of abstractions as tools in the course of scientific cognition, but we should not present the result of abstraction as a result of the investigation.

Abstraction is closest to intelligence and farthest from reality, while *deduction* is closest to reality and farthest from thought. The gap between them is bridged by *induction*, which can bring the necessary hierarchic differentiation into a knowledge of individual aspects.

In the end, however, every complete act of cognition must be based on all three methods, i.e. abstraction, induction and deduction. If the investigation is aimed at finding *Identitas* (p.169), one chiefly needs deduction, if it is aimed at A*equalitas* (p.170), one will need abstraction, and if it is aimed at *Similitudo* (p.170), one will use induction.[5]

All that can be summarized under a certain criterion may be introduced as a hierarchic level. A hierarchic level remains always an integral part of the whole

---

[5] In this connection we may be reminded of the following mottos of scholasticism:

Homo intelligens - qui habet abstractiones (who is able to distinguish the important from the neglegible)

Homo sciens - qui habet conclusiones (who knows the rules and can draw the proper conclusions)

Homo sapiens - qui habet principia (who recognizes the first causes).

system and hence it cannot be clearly defined. The mutually influencing hierarchic levels should be chosen in such a way as to allow the recognition of similarities and differences in dominance in the course of the continuous transitions.

The criterion for the appropriation of the ordering is neither the simplicity nor the precision of a model assumption, but rather its value in providing an orientation in the course of the investigation to provide an illustration of certain aspects of reality with the possibility to integrate the wealth of non-reduced observations and the results of measurements.

A hierarchically higher level is more significant for the whole system, whereas the hierarchically lowest level provides a more static framework for the dynamically better developed relations within the higher levels.

As the external conditions are changed, the highest level will "decide" in which ways the other levels will be reorganized in order to respond properly to the actions from outside. The higher levels are dynamically more active and more resistant towards changes, whereas the lower levels are more rigid and more passive. Like the river, fashioning its bed and the bed confining the river (see p. 95), the dynamic features of the higher levels shape the lower levels, which at the same time provide a kind of framework for the execution of the better developed dynamic actions of the higher levels, which respect the lower levels.

This means that a lower level is determined by the dynamic actions of the higher levels, which, in turn, are influenced by the more static boundary conditions of the lower levels, which provide a kind of framework for the dynamic actions of the higher levels.

A higher level in a hierarchy has always a much longer reaction time than a level classified as lower. This condition is necessary in order that a lower level is controlled by the dynamic variables of the higher level, and that the levels adapt themselves immediately [22]. The parts in the highest level show greater "inertness" than those in the lower levels. In this way, the hierarchically superior level appears to determine the adaptability and the resistibility of the whole system. This means that the higher hierarchic levels do not "exploit" or "enslave" the units in the lower levels - as is expressed in synergetics [19] - but provide for them the optimal conditions as they are required for their performance within the whole and for the maintenance of the structural framework for the benefit of the whole system [13].

References

1. N. S. Kurnakov, *Z. Anorg. Chem.* **88** (1914) 143.
2. H. Primas, *Chimia* **36** (1982) 293.
3. V. Gutmann and G. Resch, in *Philosophia perennis*, eds. H. D. Klein and J. Reikerstorfer, part 1, p. 382 (Peter Lang Verlag, Frankfurt, 1993).
4. V. Gutmann, *Fresenius Z. Analyt. Chem.* **337** (1990) 166.
5. J. M. Lehn, *Angew. Chem.* **100** (1988) 91 - **102** (1990) 1347.
6. C. F. v. Weizsäcker, *Die Tragweite der Wissenschaft* (Verlag Hirzel, Stuttgart, 1964).
7. J. Kepler, *De nive sex angula* (Frankfurt 1611), English translation, *The six-*

corned snow flake (Clarendon Press, Oxford 1966).

8. D. Bohm, *Wholeness and the Implicate Order*, p.173 (Routledge and Kegan Paul, London, 1980).

9. R. Spaemann and R. Löw, *Die Frage Wozu?* p.119 (Piper Verlag, München, Zürich, 1981).

10. V. Gutmann, G. Resch and E. Scheiber, *Revs. Inorg. Chem.* **11** (1991) 295.

11. A. M. McNeill, in ref. 12.

12. P. A. Weiss, in *Hierarchically Organized Systems in Theory and Practise*, ed. P. A. Weiss, (Hafner Publ. Co., New York, 1973).

13. G. Resch and V. Gutmann, *Scientific Foundations of Homeopathy*, (Barthel Publ., Berg, Germany, 1987).

14. P. A. Weiss, *The Science of Life; The Living System - a System for Living* (Futura Publ. Co., New York, 1973).

15. V. Gutmann, *Pure Appl. Chem.* **63** (1991) 1715.

16. V. Gutmann and G. Resch, *Comments in Inorg. Chem.* **1** (1982) 26.

17. C. E. Shannon, *Bell. Syst. Techn. J.* **27** (1948) 379, 623.

18. L. Bertalanffy, *General System Theory* (George Braziller, New York, 1968).

19. H. Haken, *Synergetics, an Introduction*, 3rd ed. (Springer, Berlin, 1983).

20. P. Weiss, *J. Theor. Biol.* **5** (1963) 389.

21. R. Flury, *Homeopathy and the Principle of Reality*, eds. G. Resch and M. Flury-Lemberg (M. Flury-Lemberg, Bern, 1979).

22. H. Primas, *Chemistry, Quantum Mechanics and Reductionism*, p.318, (Springer Berlin, 1983).

CHAPTER 17

SYSTEM ORGANIZATION OF LIQUID WATER

1. General Considerations

Water is unique in its enormous versatility. In particular, with regard to the functions in living bodies, liquid water is absolutely necessary, a "conditio sine qua non". In order to fulfil the enormous amount of highly specific tasks, water must have a nearly unlimited number of capabilities. Their development requires the most comprehensive system organization possible.

We should even be prepared to consider that water can adapt itself to all of the hierarchical orders of nature. Although this may sound provocative, it should not indicate just one of the possibilities borne out of our imagination, but rather express the minimum requirement for the observed actions of water in *all* resorts of nature on earth.

If we are prepared to avoid the introduction of a-priori prepositions, we ought to investigate for each aqueous solution under consideration the minimum requirements

    (i) for its existence,

    (ii) for its stability[1] and

    (iii) for its functionality.

According to these considerations, the molecules of liquid water may be grouped with respect to their significance for and within the whole system, which acts as a *unity*, or for and within one of its selected subsystems. The significance may be related to its existence, its stability and its functionality.

A group of molecules of similar significance is introduced as a *hierarchic level*. As outlined in the previous section, this introduction is the result of abstractions, as no boundaries are observed within the continuous relationships. For the description the introduction of artificial discontinuities is necessary. These are based on observable differences according to certain criteria, so as to express similar differences in dominance in the course of the continuous interactions.

From the statistical point of view the investigation of hierarchic levels is of little interest because in this approach both the similarities and the differences in qualities are eliminated and replaced by the assignment of equal numbers to unequal things in the course of the statistical procedure.

The investigation of hierarchies is, however, essential if an understanding of the qualities is sought.

---

[1] Apart from thermodynamic stabilities, thermal and chemical stabilities are considered just as well as the "abilities" to maintain metastable states under varying conditions.

It will be shown that

(i) the quality of the system is predominantly determined by the actions in the higher levels,

(ii) the quantitative data are determined by the lower levels.

Hierarchy expresses the fact that the higher levels include the abilities of the lower levels, whereas the lower levels do not include those of the higher levels, although they are influenced by them.

Hierarchic levels were proposed in 1980 for the solid state with special consideration of metastable states, such as amorphous bodies or materials treated by the actions of mechanical forces [1-4]. It has been suggested that all of the so-called defects, which are characteristic for each real solid material, are not only requirements for its very existence, but also of different hierarchical dignity [1-8].

Whereas in solid materials the hierarchic levels can be described on structural grounds, this is hardly possible for liquids because of the limited structural knowledge accessible. Because a solid may be considered as a *solid solution* and because other than structural criteria are necessary even for the characterization of the levels in the solid state, we felt encouraged to attempt a description of the qualitative changes in liquid systems by introducing hierarchic levels to liquid water [10,11] and to learn more about them in due course [12-14].

For liquid water four different groups of molecules have been described in Chapters 8 to 11, namely

(i) those near and at the phase boundaries,

(ii) those around the voids and dissolved gas molecules,

(iii) the molecules surrounding hydrophilic solutes,

(iv) all other molecules.

Each of these groups is mandatory for the existence, the stability, and the functionality of the liquid and each of them contributes in different ways to them. We suggest considering each of these groups as a *hierarchic level* [10,11], although clear-cut borderlines between the various hierarchic levels cannot be drawn, because all of them are integrated within the continuous relationships of the liquid under consideration.

It is for this reason that water molecules are not fixed in any specific level but all of the *water molecules have the ability to serve in any of the hierarchic levels*. In order to study these abilities, they must be examined in full action, i.e. in the highest hierarchic level. They lose some of these properties when they are transferred to a lower level and develop again these properties when promoted to the higher level.

Again, we wish to emphasize, that the introduction of hierarchic levels is the result of abstractions, but this should not lead to the impression that a new model is presented. Instead, the introduction of the hierarchic levels is made in order to provide help and assistance for a greater understanding of the qualitative features of the given solution under varying conditions.

2. The Boundary Areas as Highest Hierarchic Level

In establishing the hierarchy in liquid water, the molecules at and near the phase boundary have been appropriated the prime position and are proposed to represent the *highest hierarchic level that can be observed.*

It has been shown on p. 79 that the phase boundary is a requirement *for the existence and for the observability* of the liquid. In order to provide for the existence of the liquid, the boundary molecules have to be in states of high energy and of tension. As the surface energy and the surface tension vanishes at the critical point, the liquid ceases to exist.

It has also been pointed out that this state of tension may be described by the "electric double layer" which has been explained on a molecular level by the charge transfer involved in the course of the interactions of the liquid interface with its environment. It has also been emphasized that the phase boundary of the liquid belongs actually to both phases. The continuous interaction between the two phases leads to increasing negative net-charges at the donor sites (pileup effect) and to decreasing negative net-charges at the acceptor sites (spillover effect) (Chapter 4) [6].

With regard to the *stability range of the liquid* the size of its interface area as related to its volume is of importance. By decrease in drop size or by decrease in thickness of a liquid film, the ratio of interface to volume is increased and so is, therefore, the number of molecules in the highest hierarchic level.

By these changes the influence of the dominating forces at the highest hierarchic level on the whole system and hence on all of the subordinated hierarchic levels is increased and the temperature range over which a liquid is stable, is widened. The increase in dynamic properties of the whole system is reflected in the low melting point of small drops [15] and of films of low thickness [16] (p. 85) and the non-freezability of hydration layers around biomolecules down to -60°C [17] (p. 85). The highly developed properties at the interface are also evidenced by the "abnormally high" values in heat capacities of thin layers of water [18].

These highly developed dynamic properties are essential for the functionality of the liquid. Dissipative structures are being developed both in the course of the adaptability of the liquid towards changing environmental conditions and of the development of counter-actions against those influences from outside by which the maintenance of the integral configuration of the liquid might be destroyed (see p. 25 ff).

This means that at each interface an *existential combat* is taking place between the highest hierarchic levels of differently organized phases and this may lead either to their coexistence or to the annihilation of the less organized phase. Coexistence requires mutual adaptabilities and cooperativities of the interacting phases of different system organization and establishment of a dynamically maintained equilibrium.

This requires that the interface molecules are not in thermal equilibrium, but rather flexible and in positions that provide for the equilibrium conditions for the whole liquid under different environmental conditions.

The development of dissipative structures in the course of adaptation to new environmental conditions leads to local and temporal changes in composition and structure at the interface and these changes lead to certain changes in static boundary conditions in all of the subordinated levels.

This means that the redistribution of matter and energy within the liquid is controlled and regulated by the dynamic forces acting in the highest hierarchic level. All material changes at the interface, for example by adsorption, desorption, evaporation, vapour deposition cause less pronounced changes within the liquid and these processes are statistically described.

The appropriate actions of the highest hierarchic level imply another minimum condition, namely the availability of all of the information about the static and the dynamic aspects of order within the liquid at its interface, as "seen" from each point of the latter [4].

3. The Decisive Role of Voids and of Dissolved Gases

Whereas it cannot be denied that the phase boundary is a requirement for the existence of a liquid, it may appear less obvious, if not strange, to accept the presence of holes and of dissolved gases as another existential requirement for the liquid (see p. 101).

The presence of holes has been suggested on theoretical grounds long ago, namely to account for the viscosity of liquids. Holes are regarded as playing a similar role in a liquid phase as do molecules in the gas phase, namely moving about in the liquid just as do the molecules in a gas. Their presence has been well-established by the results of X-ray work on clathrates, in which the majority of the holes are found occupied by gas molecules.

Because no liquid is known which is totally free from dissolved gases, we can assume that their presence is one of the conditions for the liquid state. If this is correct, then the dissolved gas molecules ought to contribute in highly specific ways to the stability and to the functionality of each liquid.

It has been stated on p.101 that holes in a liquid seem to provide a counter pressure towards the capillary pressure. The "inner surfaces" seem to be required for the stability of the interface, which is undoubtedly a requirement for the existence of the liquid [19]. Unfortunately, the expected relationship between surface tension and hole concentration is not found. By increase in temperature the hole concentration is increased, whereas the surface tension is decreased.

It is known, however, that the stability of the holes is increased by the presence of entrapped gas molecules. Gas solubility is known to increase as the temperature is decreased. At lower temperatures, the increased surface tension is paralleled by an increase in gas solubility. The contribution of gas molecules to the stability of the liquid accounts also for the fact that under atmospheric pressure liquid water is not stable at temperatures above $+280°C$, but such stability is provided under pressure, i.e. when the gas solubility is increased.

The increased stability of the holes by entrapped gas molecules is due to their well-developed dynamical properties. The dissolved gas molecules are known to be subject to librations and rotations [20] (p. 103). Their amplitudes are confined by the flexible inner surfaces of the cavities. Mutual interactions are taking place between the entrapped gas molecules and the oscillating pattern of the surrounding liquid as mediated through the inner surface areas of the cavities, which provide a kind of flexible framework of boundary conditions and these seem to be decisive for the functionality of the gas molecules within the liquid phase.

By the mutual interactions between the molecules of the inner surfaces and the entrapped gas molecules all of them become activated. The inner boundary areas are strengthened and gain a greater influence on the lower hierarchic levels.

The motions of the gas molecules are reflected in their high heat capacities (p.103), which have been calculated to be even higher than in the gas phase at the same temperature [21]. This shows that the dissolved gas molecules provide for the presence of certain aspects of the gaseous state within the liquid phase.

The modifications of the oscillation pattern of the liquid due to interactions with the gas molecules lead to a "loosening" of the liquid structure which extends, in principle, over the whole solution.

At the same time the oscillating gas molecules are modified in their rotational pattern according to the conditions of the oscillating pattern of the liquid, until a certain "synchronization" is established. It may be supposed, however, that the synchronization does not involve metronomic frequencies, but rather movements of regular successions within a certain frequency pattern, like a "kind of melody" that is characteristic for the quality of the system under consideration.

By these mutual interactions a kind of "relais" appears to be established by which certain weak actions may be amplified and certain strong actions distributed in highly specific ways over the whole system. The motions of the dissolved gas molecules are confined by the boundary conditions of the holes of the liquid and appear in this way to regulate and to control all other molecules of the liquid except those of the highest hierarchic level.

This means that the *dissolved gas molecules have the remarkable ability to take over structural information from the solution structure and to integrate it and to preserve it dynamically within their oscillations, in harmony with those of the solution* and still under the control of the forces acting in the highest hierarchic level.

As the oscillating gas molecules appear to maintain dynamically the structural information of the whole solution, they must be assigned a high hierarchic significance with regard to the preservation of the integral configuration and functionality of the whole liquid, i.e. as serving the hierarchic level that is immediately subordinated to that of the phase boundaries and superordinated to all other levels of the liquid.

Both of these higher levels determine the quality of the liquid system, whereas all the other molecules, serving on subordinated hierarchic levels, determine the quantitative aspects found for the solution. As the physical chemistry of solutions is concerned with quantitative aspects rather than with qualities, the role of the higher hierarchic levels is of little importance.

This leads to new questions which have not so far been asked: in what ways can the structural features imposed on the solution by hydrophilic solutes reach the inner surfaces and in what ways can the static aspects of structure be transformed into oscillating properties and maintained dynamically by them?

4. The Role of Hydrophilic Solutes

It has been mentioned on p.109 that another existential requirement for liquid water is the presence of hydrophilic solutes. Hydrated hydrogen ions and hydrated hydroxide ions are bound to be present due to the self-ionization equilibrium and it is unlikely that the last traces of other hydrophilic solutes can be removed.

As all of these solutes contribute to the structural and energetic differentiation of the liquid, they are hierarchically more significant than the vast amount of bulk solvent molecules.

With regard to the stability and functionality of liquid water, the hydrophilic solutes are, however, of lower hierarchic significance than the dissolved gas molecules. It has been shown in Table 18 on p.104 that addition of hydrophobic and of hydrophilic solutes to water leads to different changes in liquid properties.

It has also been shown that *hydrophobic solutes are more "independent" of water and dynamically more active than hydrophilic solutes.* The former lead to a loosening of the water structure, whereas the latter lead to a tightening of the water structure.

These differences in structural changes reflect the different significance of hydrophobic and hydrophilic solutes with regard to their functionality within the liquid. As outlined in the preceding section, the former help to preserve dynamically the quality of the solution. On the other hand, the hydrophilic solutes lead to structural changes of the solution by which they are themselves modified. These modifications are important to the chemist in his choice of the most suitable solvent for a particular reaction.

Whereas the hydrophobic solutes contribute in the first place to the maintenance of the quality and functionality of the liquid, the hydrophilic solutes lead to observable and measurable changes in structural and thermodynamic properties which can be expressed in quantitative ways. Since a *quality is required for a quantity to be measured*, the hydrophobic solutes are assigned a higher significance for the whole system than the hydrophilic solutes and their surrounding molecules. The hydrophilic solutes appear subordinated to the hydrophobic solutes but superordinate to the molecules which are responsible for the structural framework of the liquid.

This superordination is implied in all studies of solution chemistry, which are in the first place concerned with the properties and reactivities of the dissolved species.

It has already been shown in Chapters 13 to 15 that studies in non-aqueous solutions are confined to the influence of the bulk medium on structures and reactivities of the solutes, as the study of the dominating influence of the solutes species on the medium properties appears to be too difficult from the experimental point of view.

The superordination of the solutes to the vast amount of bulk solvent molecules is, in fact, implied in all thermodynamic, kinetic, structural, and spectroscopic studies concerning solute properties and reactions in solution (such as complex formation, substitution reactions, redox reactions).

5. The Role of all other Solvent Molecules

The vast amount of molecules are somewhat remote from the phase boundary, from the gas molecules and from the hydrophilic solutes. These appear to serve the lowest hierarchic level of the solution. This is, however, also a requirement for the very existence of the liquid.

Although its constituent molecules have lower energies per part and less developed dynamic properties than the molecules in the higher levels, the lowest level has the greatest energy storage capability because of the large number of molecules (Fig. 57 on p. 193).

This level provides the structural framework for the whole system and this can be shaped and modified according to the requirements of the superordinated levels. At the same time the lowest level is respected and used by the higher levels. Each of the molecules in the lowest level is least affected by the actions of the higher levels, but "knowledge" of the properties of the molecules in the lowest level is inadequate to understand the system or to make conclusions about their properties when they are promoted to a higher level.

The driving forces for the migrations in all of the hierarchic levels of the liquid are gradients in chemical potential. As migrations are taking place in microregions, new gradients in chemical potentials are produced in other microregions, so that further migrations are bound to follow in due course.

This means that the macroscopic state as characterized by structural and thermodynamic data in the lower levels represents a great number of dynamically co-existing (fluctuating) microscopic states which cannot be distinguished from each other [22]. By changes in microregions, *a characteristic motion pattern is maintained which is independent of the actual positions of the molecules*. What remains constant is neither the local arrangement of the molecules or ions, nor the local analytical composition, but rather the regularities of the dynamic character of the whole system.

By the statistical description of these processes as diffusion, the influence of the temperature on the motion pattern with its regularities is usually considered, but little attention is paid to other influences such as the actions of fields, of irradiation, of mechanical forces and even of the chemical properties of the environment. In this way the temperature of a liquid is usually related to "thermal vibrations". However these actually represent a complicated dynamical pattern with a character of its own that reflects the total energetic influence on the liquid and not only that of its temperature.

6. Illustration of the System Organization

All of the above considerations may be illustrated by Fig. 56 by means of a truncated pyramid [6,11]. No apex is shown for the pyramid because this would represent the highest hierarchic level which is, however, sensually not perceptible. It is therefore impossible to investigate this highest domain, although it should be anticipated that this dominates and controls all of the observable hierarchic levels shown in the truncated pyramid (Fig. 56). Any attempt to gain access to this highest level would require substantial changes within the whole system with simultaneous disturbance of the organization to be investigated.

The higher the level, the greater is its hierarchic significance for the maintenance of the liquid under different conditions. The dynamic properties of the molecules are most pronounced in the highest level and the energy content and the energy storage capability per molecule is greatest (Fig. 57).

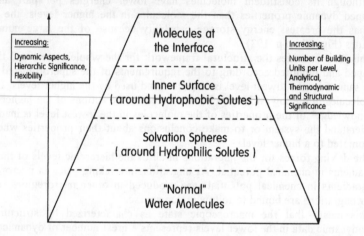

Increasing:
Dynamic Aspects,
Hierarchic Significance,
Flexibility

Molecules at
the Interface

Inner Surface
(around Hydrophobic Solutes)

Solvation Spheres
(around Hydrophilic Solutes)

"Normal"
Water Molecules

Increasing:
Number of Building
Units per Level,
Analytical,
Thermodynamic
and Structural
Significance

Fig. 56. Illustration of the system organization of liquid water by means of a truncated pyramid.

It has been convincingly demonstrated for the solid state that the interface particles may gain or lose considerable amounts of energy without losing their dominating properties. In the course of a cooling process they retain more energy from the state of higher temperature than other particles [23,24]. This means that they remain dynamically more active than would be expected from purely thermodynamic considerations. The same seems to be true for water which is obtained by rapid cooling in a non-crystalline (glass-like form), i.e. the so-called "vitrified water" [25,26]. In this metastable state the whole system is in a more highly differentiated and dynamically better developed state than in the crystalline state of ice obtained by a normal cooling procedure.

The lower the level, the less pronounced are the dynamic aspects, the smaller the flexibility and the smaller the energy content per part. Because the number of molecules is greater, the lower the level, the lowest level contains the greatest fraction and the highest level the smallest fraction of the total energy of the system (Fig. 57).

For the redistribution of energy the following rules have been formulated [11]:

(i) The amount of energy that can be adsorbed by the liquid is greater the lower the temperature and the greater the surface area as related to the volume, i.e. in small drops, in thin layers, and in colloidal systems.

(ii) Unless very small drops are considered, the increase in energy within a given level will be greater the lower the level, but the energy increase per molecule will be greatest in the highest and smallest in the lowest level (Fig. 57).

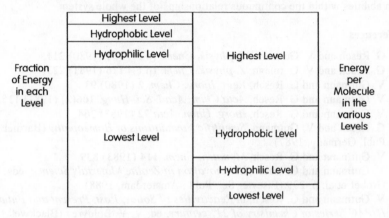

Fig. 57. Illustration of the energy distribution within liquid water.

For this reason the properties of the molecules in the lower levels contribute most decisively to the statistical results. For the assignment of quantitative data on thermodynamic, kinetic, spectroscopic, and structural properties it is sufficient to consider the two lower levels, i.e. that of the hydrophilic solutes and of the bulk solvent molecules (see also Chapter 18).

The properties of the molecules in the highest levels contribute little to the statistical data and this may be one of the reasons why the highest levels are not considered in physico-chemical studies of solutions unless special problems are treated, such as interface phenomena or gas solubility. *Modern solution chemistry is therefore concerned only with the two lowest hierarchic levels by which quantitative aspects are determined.*

The higher hierarchic levels are directly observable, more complex and more flexible than the lower levels which are less differentiated and less sensible towards changes. The highest levels appear mainly dynamically active, resistant against forces and load, and responsible for the quality and individual characteristic of the liquid. The lowest levels are more passive, more rigid, and decisive for quantitative characterizations.

It is, however, important to note that *all of the levels are necessary for the existence of the system and for its understanding,* as each of them serves certain objectives and goals for the whole liquid. As the external conditions are changing, the highest level will "decide" in which way all of the levels will be reorganized in order to respond properly through their concerted interactions for the maintenance of the chief characteristics of the system.

As all of the molecules are continuously interacting with each other, and as the whole system is responding to all changes from outside, *none of them is fixed in a certain level.* The actual properties of the molecules depend on the role they have to perform according to the requirements of the whole system and according to their own abilities within the continuous relationships of the whole system.

References

1.   G. Resch and V. Gutmann, *Z. physik. Chem.* (n.F.) **121** (1980) 211.
2.   G. Resch and V. Gutmann, *Z. physik. Chem.* (n.F.) **126** (1981) 223.
3.   V. Gutmann and G. Resch, *Revs. Inorg. Chem.* **2** (1980) 93.
4.   V. Gutmann and G. Resch, *Acta Chim. Acad. Sci. Hung.* **106**(2) (1981) 115.
5.   V. Gutmann and G. Resch, *Inorg. Chim. Acta* **72** (1983) 264.
6.   G. Resch and V. Gutmann, *Scientific Foundations of Homeopathy* (Barthel Publ. Germany, 1987).
7.   V. Gutmann and G. Resch, *Monatsh. Chem.* **114** (1983) 839.
8.   V. Gutmann and G. Resch, in *Innovation in Zeolite Materials Science*, eds. P. J. Grobet et al. p. 239 (Elsevier Sci. Publ., Amsterdam, 1988).
9.   V. Gutmann and G. Resch, in *Reactivities of Solids, Past, Present and Future, IUPAC Series of Chemistry of 21st century*, ed. V. V. Boldyrev (Blackwell, Oxford, to be published).
10. G. Resch and V. Gutmann, in *Advances in Solution Chemistry*, eds. I. Bertini et al, p.1, (Plenum Press, New York, 1981).
11. V. Gutmann and G. Resch, *Pure Appl. Chem.* **53** (1981) 1447.
12. V. Gutmann, E. Scheiber and G. Resch, *Monatsh. Chem.* **120** (1989) 671.
13. V. Gutmann, G. Resch and E. Scheiber, *Revs. Inorg. Chem.* **11** (1991) 295.
14. V. Gutmann, *Pure Appl. Chem.* **63** (1991) 1715.
15. D. M. Anderson, *Nature* **216** (1967) 563 - *J. Coll. Interface Sci.* **25** (1967) 174.
16. A. Tsugita, T. Tarei, M. Chikazawa and T. Kanazawa, *Langmuir* **6** (1990) 1461.
17. A. D. Buckingham, *Disc. Farad. Soc.* **24** (1957) 151.
18. M. Elzler and J. J. Conners, *Langmuir* **6** (1990) 1250.
19. T. Emi and J. O'M. Bockris, *J. Phys. Chem.* **74** (1970) 159.
20. J. M. Van der Waals and J. C. Platteeuw, in *Advances in Chemical Physics*, ed. I. Prigogine, **2**, 1, (Interscience Publ., New York, 1959).
21. E. Wilhelm, R. Battino and R. J. Wilcock, *Chem. Revs.* **72** (1972) 211.
22. P. Glansdorff and I. Prigogine, *Thermodynamic Theory of Structure, Stability and Fluctuations* (Wiley-Intersci. London, New York, 1961).
23. V. Gutmann, G. Resch, R. Kratz and H. Schauer, *Z. anorg. allg. Chem.* **491** (1982) 95.
24. V. Gutmann, G. Resch, R. Kratz and H. Schauer, *Monatsh. Chem.* **115** (1984) 559.
25. P. Brügeller and E. Mayer, *Nature* **298** (1982) 715.
26. J. Dubochet and A. W. McDowell, *J. Microsc.* **124** (1981) RP3-RP4.

CHAPTER 18

CHANGES IN ORGANIZATION OF LIQUID WATER

1. Changes in the Temperature Range between 0°C and 100°C

A change in temperature leads to a change in energy content and to changes in all of its properties, such as density, viscosity, conductivity, vapour pressure. As the temperature is lowered the surface energy is increased and the vapour pressure is decreased. The dynamic properties are decreased in all levels, but to a lesser extent in the highest level.

Due to increased gas solubility, more energy can be retained in the level that is immediately subordinated to the highest level. This leads to a loosening of the water structure which is, however, overcompensated by the loss of energy in the lowest level so that the static boundary conditions are strengthened. Under atmospheric pressure at 0°C the dynamic properties are no longer suitable for the maintenance of the liquid state and the molecules become arranged within the more rigid crystal lattice characteristic of ice.

Although the molecules in the higher hierarchic levels are in higher energy states than those in the lowest levels, ice floats to the surface due to the density maximum of liquid water at +4°C and to the additional decrease in density by the freezing process. The ice at the surface is pushed out of the liquid and provides a certain "shelter" for the liquid due to its well-developed static boundary conditions of low thermal conductivity.

When the temperature is raised again, the interface area is the first to melt, but the increase in energy is mainly passed on to the lower levels. The energy content is increased in all hierarchic levels, although to the greatest extent in the lowest level. The decrease in density above +4°C, the increased conductivity, and the slight changes in the infrared spectra reflect the increased dynamization in the lower levels. They seem to be more reluctant to retain much energy, as is indicated by the slight decrease in heat capacity below +37.5°C.

This may be related to the decrease in gas solubility, (which implies a decrease in the number of "synchronizing centres"), whereas the number of unoccupied cavities is progressively increased as the temperature is raised; the water structure is thus considerably loosened.

As the temperature is further raised, the dynamic features are further improved in the highest hierarchic level as is seen from the decrease in surface tension and the increase in vapour pressure.

The increase in dynamic properties of all of the hierarchic levels leads to greater reactivity in its existential combat with the environment, but at the same time provides for a certain weakness due to the weakening of the static boundary conditions. When, under atmospheric pressure, the temperature of 100°C is reached, the minimum of the static boundary conditions as required for the existence of the liquid state is no longer provided and the liquid is evaporated.

The system organization may be considered as optimal when a certain "balance" between static boundary conditions and dissipative actions is reached. As the increase of the former is reflected in decreasing values for the heat capacity, the minimum found for liquid water at +37.5°C may be an indication of such well-balanced conditions.

2. Supercooled Water

Under selected conditions purified liquid water is found to exist down to a temperature of -44°C ("supercooled water") and well above +100°C (superheated water"). The increased resistance to lower and higher temperatures respectively requires an improved system organization. We have tried to learn about this improvement by considering carefully

(i) all of the conditions under which liquid water is successful in its existential combat with the environment below 0°C and

(ii) all of its properties under these conditions.

In this way we have attempted to follow the demand made on p. 171 to gain knowledge by starting the investigation under complex conditions.

Although it is well-known that by addition of hydrophilic solutes the freezing point is lowered and the boiling point is elevated, it must be pointed out that the liquid range is extended to a much wider temperature range *provided that liquid water of high purity is cooled* under conditions as outlined on p. 58 ff.

Its highly developed dynamic properties are partly retained at low temperatures partly due to special conditions of the cooling mode, namely the quenching method. Numerous examples can be quoted for this effect, such as the highly developed dynamic properties of films produced by rapid vapour quenching procedures. For example, rapidly quenched cobalt is obtained at room temperature in the structure of the high temperature phase [1]. A better known example is that of the *Martensitic* transition which is of great importance in steel technology.

Likewise, rapidly quenched water may be obtained in a glass-like state, known as "vitrified" water [2,3]. In this state, water is more differentiated and dynamically better developed than crystalline ice. In order to obtain supercooled rather than vitrified water the temperature difference for the quenching process and the cooling rates must be properly adjusted.

Because the energy at the interface is dynamically maintained by the oscillations of the interface molecules, it is clear that the dynamical features of the highest hierarchic level are improved as the number of molecules in this level are increased by increase in interface area [4]. This is in agreement with the observed fact that a lower temperature of the liquid is reached where the diameter of the capillaries or the size of the droplets is decreased (p. 59 and p.60).

The enormous improvement in dynamic properties attained by increase in the ratio of surface area to volume may be illustrated by the properties of films of gold of thickness below the μm-region [5]. Such films of gold are known to flow together at temperatures many hundred degrees below its melting point [6]. It has been shown that under certain conditions an optimal system organization is reached [5] (p.203).

The influence of the environment has been expressed by *Angell* [7] as follows: "It has become quite clear, however, that crystallisation of ice as it is normally observed is not a property of water itself, but rather a function of solid surfaces, particulate or otherwise, with which it is in contact".

It can be seen from Fig. 9 on p.60 that lower temperatures can be obtained if the capillaries have been provided with a hydrophobic coating (see Chapter 19). In extremely small droplets in a hydrophobic atmosphere, the liquid can both retain its energy content while the dissolution of gas molecules can readily take place. The strengthening of the system organization under these conditions is reflected in the low values for the surface tension and in the high values for the vapour pressure and these properties correspond to those of bulk water at a much higher temperature.

These properties of small drops[1] are additionally stabilised by increased gas content, which leads to a definite strengthening of the level which is immediately subordinated to the highest hierarchic level (p.100). The preservation of dynamical features is progressively improved as the gas content is increased due to the lowering in temperature.

The improvement of dynamic features due to increased gas content is reflected in the progressive increase in the values for the specific heat of supercooled water as the temperature is lowered. This "abnormal" and dramatic increase in heat capacity by lowering the temperature (Fig. 10 on p.61) is related to the increased gas concentration which helps to increase the dynamic aspects (apparently to compensate the strengthened static boundary conditions). It is, therefore, *not anomalous, but rather a requirement for the existence of supercooled water.*

The highly improved dynamic properties in the higher hierarchic levels have a profound influence on the dynamization of the lower levels. They provide for a loosening of the "water structure" which is also reflected in the decrease in density (Fig. 11 on p.62) and the increase in compressibility (Fig. 12 on p.62).

The considerable improvement in differentiation and in dynamic properties in the higher hierarchic levels allows the exercise of their full influence on the dynamization of the lowest level, as long as no additional "centres of excellence" are provided by the presence of hydrophilic solutes (see p.199 ff), as these provoke local changes within this hierarchic level.

The system organization of supercooled water seems optimally developed when all of the "instructions" from the higher levels reach the lowest level undisturbed. In other words: the system organization of liquid water is improved

(i) as the higher hierarchic levels are more differentiated and
(ii) as the differentiation in the lower levels remains as small as possible.

With regard to the number of molecules in the various levels we refer to Fig. 58 which shows that the well developed system organization of liquid water is characterised by the increase in the number of molecules from the highest to the

---

[1] The surface tension of a droplet is known to decrease with increasing curvature [8].

lowest level with a broad basement in the large number of molecules in the lowest level.

Fig. 58 shows that in supercooled water the number of building units in the higher levels is appreciably higher, while the number of molecules in the lower levels remains about the same as in pure water. The dashed lines in Fig. 58 show a greater slope for supercooled water than for liquid water and this illustrates the improvement of system organization in supercooled water.

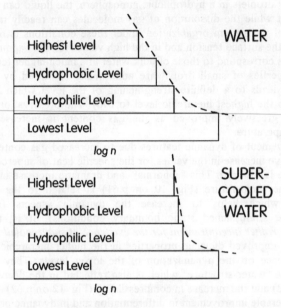

Fig. 58. Influence of the number of molecules n in the various levels on the system organization of liquid water and of supercooled water. The improvement of the system organization in supercooled water is indicated by the greater slope of the upper part and the smaller slope of the lower part of the dashed line.

The free enthalpy difference and the entropy difference between supercooled water and ice is greater for water cooled in emulsions than in capillaries. As shown in Fig. 14 on p.64, the temperature coefficient for the differences in free energies in emulsions and in capillaries is greatest in the temperature range -20°C to -30°C. In the same temperature range the greatest differences are also found in entropy values (Fig. 15). It has been suggested on these grounds that the system organization is better developed in water cooled in emulsions than in capillaries and that the system organization is best developed in the said temperature range between -20°C and -30°C [4].

3. Water in Thin Layers

The consideration of supercooled water obtained in capillaries has shown that the system organization is improved in thin layers. This is in agreement with the higher vapour pressure, the high values in heat capacity (p. 83) and the lower surface tension and density at the same temperature. Such properties are found for bulk water at a higher temperature. The increase in dynamic aspects in extremely thin layers is reflected in the low melting points (Fig. 18 on p. 83) and the enhanced reactivities and solubilities.

For example, water in capillary tubes attacks the walls and leads to dissolution of silicic acid to a much greater extent than is expected from equilibrium considerations. This explains the formation of "flints" in chalk that follow from the dissolution of quarz in fissures in surrounding quarz based minerals.

In the existential combat at the phase boundary the liquid exhibits a highly developed system organization, which allows it to integrate matter, energy and information to a greater extent than in the less developed state of organization as is characteristic for bulk water.

4. Actions of Hydrophilic Solutes

It is well-known that by addition of hydrophilic solutes the temperature range for the stability of the liquid phase is also extended. Vapour pressure and freezing point are decreased and the boiling point is increased.

The dissolution of hydrophilic solutes leads to an increase in the number of molecules in the hydrophilic level which is subordinated to the higher levels and superordinated to the lowest hierarchic level. Because the solutes move freely within the liquid, they contribute to increased dynamic properties in the lower levels.

The rearrangement of the water molecules within the hydration layers leads, however, to local increases in density and to a tightening of the water structure. For this improvement in static boundary conditions in the lower levels, the solute molecules or ions provide a kind of "centres of excellence" within the liquid. Their influence cannot end somewhere in the liquid and hence this leads to a tightening of the structure even in the highest hierarchic level. This is clearly reflected in the lowering of vapour pressure and in the increase in surface tension.

This further shows that structural information from the lower levels is made available to the highest hierarchic level, which becomes dynamically less active in the presence of hydrophilic solutes. The same is true for the level that is immediately subordinated to the highest level. The dissolution of hydrophilic solutes is connected with the "salting out " of dissolved gases and hence with the destabilization of the voids.

With regard to the static structural aspects, the solution corresponds to pure water at a lower temperature, but with regard to the decrease in gas concentration it corresponds to water at a higher temperature. The hydrophobic level becomes less differentiated by a decrease in gas concentration, whereas the subordinated levels become increasingly differentiated as the solute concentration is increased. This increased differentiation is reflected in the broadening of the lines in the relaxation spectra and the effects of orientation polarization [9].

As the temperature is raised much of the energy gained by the solution seems to be stored within the most differentiated hierarchic level, i.e. the level of the hydrophilic solutes as the number of solute molecules has been increased. Part of this energy is passed on to the lowest level, which, at the same time, is somewhat dynamized by the motions of the solute molecules within the liquid.

Under these circumstances the dominating influence of the higher levels on the whole liquid is weaker than in pure water, whereas the influence of a lower level, namely that of the hydrophilic solutes, is strengthened.

The great significance of the hydrophilic level in a reasonably concentrated solution is well demonstrated by the rather singular results obtained for the rate coefficients for the decomposition of $[FeL_5^{3+}Fe(tmphen)_3^{2+}]$ in dimethylsulfoxide [10]. The results presented in Fig. 47 show that the rate coefficient is nearly independent of temperature and this implies that the changes in energy of the solution due to variation in temperature are redistributed over all of the hierarchic levels in such ways that the energy content of the solute level remains unchanged.

As the dominating influence of the highest hierarchic levels on the whole liquid is weakened and that of the lower levels strengthened, the system is less adaptable towards changes in environment and hence its system organization less developed than that of pure water.

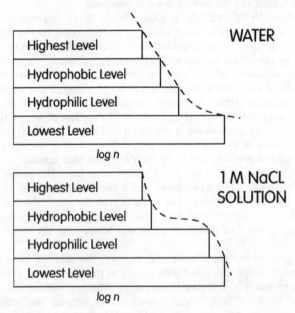

Fig. 59. Plot of the logarithm of the number of molecules n in the various hierarchic levels for water and for a 1 molar sodium chloride solution.

The differences in system organization between pure water and a 1 molar sodium chloride solution are illustrated in Fig. 59, which has been designed in analogy to Fig. 58.

As in a 1 molar sodium chloride solution the number of solute molecules is considerably higher and the number of gas molecules lower than in pure water at the same temperature, the dashed line in Fig. 59 has a different shape than that for pure water. It indicates the small influence of the highest hierarchic levels (which determine the quality of a solution) and the great influence of the lower levels which determine the quantitative aspects of the solution. The system organization of the solution appears to be less developed than that of pure water.

From this point of view the following restrictions in physical chemistry in solutions become obvious:

(i) Most investigations are devoted to solutions of concentrations above 10^{-4}mol/l. Under these conditions the lowest hierarchic level seems to be highly influenced by the great number of solute molecules or ions, so that the influence from the highest hierarchic levels is no longer fully effective. The system is less adaptable to the environment and therefore *not optimally organized* [13]. In addition, the experiments are performed under selected environmental conditions, so that natural influences of the environment are not taken into consideration.

(ii) Under all of these conditions the *exclusive consideration of the two lowest levels is justified only for the quantitative description of the solution* and hence for the determination of thermodynamic, kinetic and structural data.

5. Electrode - Electrolyte Interfaces

It has been pointed out that an existential combat between the two phases is always taking place at a phase boundary where the highest hierarchic levels of the two phases are interacting. This combat may lead to

(i) an equilibrium and to the coexistence of both phases or

(ii) to the annihilation of the less organized phase. *Eigen* and *Winkler* [12] point to the natural necessity that the system of greater stability must defeat that of lower stability, i.e. that nature seems to know only victor and loser, and all that remains of the latter may be the memory of the former to the combat, such as the wounds received in the course of the battle [13].

The enormous influence of slight changes in interface properties is indicated by the potential differences measured between two copper electrodes taken from the same copper rod placed in the same copper sulphate solution [14]. The potential differences are found to be between 0.1 mV and 10 mV and they are further increased when the interfaces of the electrodes have been treated in different ways. This effect is also found by measuring the redox potential of a copper rod towards a given reference electrode. The potential differences are smaller when oxygen is excluded and they increase as more oxygen is admitted.

The "passivation" by the admission of oxygen seems to be due to the presence of a semiconducting selvage layer at the copper electrode [15] and to a strengthening of the highest hierarchic level [16]. This is frequently considered as a manifestation of disorder, as illustrated by the following statement [17]: "The worst case for a metal surface is represented by the interactions with oxygen" (actually it is the worst case for an attempt of an idealized description and hence for the investigator rather than for the metal!).

The redox potential of the more differentiated and dynamically more active copper surface in the presence of oxygen is found at less negative potential values (Fig. 60). The presence of oxygen provides for the passivation of the metal in moderately reactive solutions. For the copper system under consideration passivity is maintained above a pH-value of two, whereas at lower pH-values the passivated copper electrode becomes more reactive [14,18].

Under these circumstances the liquid becomes more differentiated and more reactive towards the more differentiated (passivated) than to the less differentiated (non-passivated) copper surface. The more differentiated solid yields more readily to the highly differentiated solution and is known to show higher resistance towards the less differentiated solution at pH > 2. The improved surface dynamics due to passivation (the presence of mobile oxygen in the surface) provides greater resistance to moderately differentiated solutions and greater vulnerability towards better organized solutions.

Fig. 60 shows that at pH = 2 a point of intersection is found for the potential curves for differently treated copper rods. At this point the potential difference and the reactivity of the electrode are independent of the history and of the actual state of surface reconstruction and the conditions for the thermodynamic equilibrium are perfectly fulfilled. Both the solid and the liquid phase show optimal abilities to preserve their configuration and functionality independent of history (indicating well-balanced relations between static boundary conditions and dissipative features).

Under these conditions the whole system is most adaptable towards changing conditions with optimal preservation of its chief characteristics and the equilibrium of a system is fully defined by the variables of state. The whole system may be considered to be *optimally organized*.

States of optimal system organization are obviously maintained in the course of heterogeneous catalysis. The static boundary conditions of the highly differentiated catalyst are maintained dynamically and promote in these ways reactions in the adjacent phase without alterations of the catalyst itself. This implies that well-balanced relations exist between the static boundary conditions of the catalyst and the dissipative actions in the adjacent phase.

Another example of a highly developed system organization is provided at the transition point found in the course of a phase transition in the solid state. The transition curves for samples of different mechanical treatment or mode of preparation show a point of intersection where the specific heat and the entropy changes show maximum values. These "states" are independent of the history of the material and are characterized by *optimal* penetration of both phases characterized by a certain "balance" between static boundary conditions and dissipative actions [19].

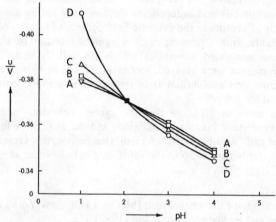

Fig. 60. Potential differences between copper electrodes of different pre-treatment and a Hg/HgSO$_4$ reference electrode as a function of the pH-value of a 10^{-4} M CuSO$_4$ solution in water.
A: Electrode etched under argon, no air admitted at all,
B: Electrode etched under argon, placed under air into the electrolyte, which had been freed from oxygen.
C: Electrode rubbed with emery under nitrogen and without etching placed in the electrolyte under air which had been freed from oxygen.
D: Electrode rubbed with emery, left in the electrolyte saturated with air for 12 hours before measurement.

States of optimal system organization have also been found for extremely thin films of gold [5], where transitions have been found to occur into a "melt-like state" at temperatures as low as 200°C. For certain conditions a "state" is found where neither expansion nor contraction occurs irrespective of the load (within certain limitations). The enormous versatility of membranes will be discussed in the following Chapter.

6. Static Aspects of the System Organization

The role of the static boundary conditions for the dissipative actions within the oscillatory "network" may be illustrated by a spider web [20]. This is strongest at the boundaries, but the breaking of a single thread effects the structure of all of the rest and hence a certain change in the static boundary conditions is produced. If a given thread were neatly stretched within the elastic network, each one of the other threads seems to "know" its own contributions to the total restoration process.

For a network of countless threads which can be distorted in an enormous number of ways, an outstandingly high "intelligence" on the parts of the components may be implied (see p. 37 and p. 231). The following comments have been made by *Weiss*[20]: "A slight switch of viewpoint will make clear how the example of an elastic net can serve as a simplified model for "system behaviour". By viewing the nodal points of the elastic net as sites of separate discrete subunits, formerly visualized as autonomous and independent entities, and letting the elastic threads stand for symbols of vectors in the dynamic field pattern of forces and "interactions" among the subunits, one attains at least a symbolic image of the conservative features of those integrated superunits which we call "systems". The case of networks whose meshes were severed, moreover, serves to illustrate how a system can transform from one equilibrium state into another without losing its systemic unity, identity and integrity."

It has been mentioned before that the system behaviour seems to be best performed when the static boundary conditions and the dissipative features are both highly developed and "well-balanced". As indicated before, this seems to be the case in supercooled water between -20°C and -30°C and in bulk water at +37.5°C.

References

1. K. L. Chopra, *Thin Film Phenomena* (McGraw Hill, New York, 1971).
2. P. Brüggeler and E. Mayer, *Nature* **298** (1982) 715.
3. J. Dubochet and A. W. McDowell, *J. Microsc.* **124** (1981) RP3 - RP4.
4. V. Gutmann, E. Scheiber and G. Resch, *Monatsh. Chem.* **120** (1989) 671.
5. V. Gutmann and G. Resch, *Monatsh. Chem.* **114** (1983) 839.
6. G. Tammann and W. Boehme, *Ann. Phys.* **12** (1932) 820 - I. Sawai and M. Nishida, *Z. Anorg. Allg. Chem.* **190** (1930) 735.
7. C. A. Angell, in *Water, a Comprehensive Treatise*, ed. F. Franks, Vol 7, p. 1, (Plenum Press, 1980).
8. R. Defay, I. Prigogine, O. Bellmans and D. M. Everett, *Surface Tension and Adsorption* (Wiley, New York, 1966).
9. J. Barthel, F. Schmithals and H. Behret, *Z. Phys. Chem.* (n.F.) **71** (1970) 115.
10. R. Schmid, R. W. Soukup, M. K. Aresteh and V. Gutmann, *Inorg. Chim. Acta* **73** (1983) 21.
11. V. Gutmann and G. Resch, *Chim. Oggi* **10** (May 1982) 15.
12. M. Eigen and R. Winkler, *Das Spiel*, (Piper, München, Zürich, 1975).
13. G. Resch and V. Gutmann, *Scientific Foundations of Homeopathy*, (Barthel Publ., Germany, 1987).
14. W. Linert, K. Stiglbrunner and V. Gutmann, *Monatsh. Chem.* **116** (1985) 1263.
15. V. Gutmann, G. Resch, R. Kratz and H. Schauer, *Monatsh. Chem.* **115** (1984) 551.
16. V. Gutmann and G. Resch, *Atti Accad. Peloritana, Classe I*, **57** (1989) 255.
17. K. Christmann, *Z. Phys. Chem.* (n.F.) **154** (1987) 145.
18. V. Gutmann, G. Resch, W. Kantner and W. Linert, *Monatsh. Chem.* **120** (1989) 11.
19. V. Gutmann and G. Resch, *Inorg. Chim. Acta* **72** (1983) 269.
20. P. A. Weiss, *Hierarchically Organized Systems in Theory and Practise* (Hafner Publ. Co., New York, 1971).

CHAPTER 19

WATER WITHIN THE HUMAN BODY

1. The Human Organism

In persuing the aim of learning about the potentialities of liquid water, we shall follow the requirement expressed on p. 173, namely to investigate the action of water under the most complex conditions in the physical world, i.e. within the human body. We meet man in full possession of his faculties, as an undivided *unity of soul and body,* despite the complexity and the variety of his functions.

Since the human body is part of the physical world, its system organization is closely related to that of nature, both animate and inanimate. We can distinguish spheres that correspond to the material aspects of the finality of the body and the different functions according to dynamically ordered relationships which exist apart from observable material conditions.

The human body contains more than 100 trillion cells. Each of them serves a special purpose, but there is a basic one-ness of all of them originating from one cell, the fertilized human ovum. It consists of a nucleus, a cell membrane and, between them, connecting cytoplasm.

The embryologist *Blechschmidt* [1,2] has shown that the individuality of the human being is already existent in the fertilized ovum and preserved until death. The first organ that develops is the brain, which in the adult contains ten thousand million nerve cells organized like a switchboard in order to serve as an important instrument for the process of perception. It has a higher hierarchic significance than the inner parts and "leads" the development from the start. In the course of ontogenesis, certain cells "take over" certain functions and functional units (the organs) are created which are made up of cells and each of them serving a particular purpose.

No cell should be considered as an independent or rigid unity, but rather as a momentary aspect of spatially ordered metabolic movement. The same is valid for cell aggregations (tissues), for tissue aggregations (organs) and for the whole organism at any stage of development.

No organ has been found to be functionless in any phase of its development. Every part of the body contributes to the formation of the total organism throughout all of the embryonic stages.

The development of the heart is a consequence of early activity of the brain. Later the kidneys are formed which appear to function very early on as excretory organs. The surface ectoderm is expanded by rapid growth of the brain and spinal cord. The formation of the skull, of the spine, and of the skeleton as structural framework, bearing mainly the conservative aspects of order, are preceded by "densations". All phases of development are movements that are directed to a certain goal. Each human being has his own finality, and therefore cannot be really understood by modern scientific methods.

The formative functions require the pre-existence of life and the implication of a soul. The human being in its concrete individuality is subject to its finality and therefore more than can be grasped by scientific methods.

A few examples may illustrate the enormous adaptability of the human body:

(i) Muscle is caused to build more muscle in response to increased load.

(ii) Vessels widen with increased blood flow and shrink with reduced oxygen pressure.

(iii) Breathing is speeded up or slowed down depending on oxygen demand.

(iv) The reciprocal fitness among the various organs is demonstrated by hormonal and neural correlation, by the matching of muscles and skeleton.

(v) Within all aggregates of tissue there are interdependencies between tissue cells and blood.

(vi) Within cells we note harmonious coordination of molecular processes, such as the selective relations of enzymes to substrates or the orderly sequences of steps in the metabolic cycle.

It is, however, not possible to understand the functions of a cell by considering a "molecular control of cellular activity" as the molecular activities appear to be controlled by the cell and its function within the whole organism. The article by *Weiss* entitled "From Cell to Molecule" appeared, however, in the book entitled "The Molecular Control of Cellular Activity" [3]. Subsequent publications followed in due course [4-8]. The following quotation has been taken from his first article [3]:

"The story of molecular control of cellular activities is bound to remain fragmentary and incomplete unless it is matched by knowledge of what makes a cell the unit that it is, namely the cellular control of molecular activities... We have to acknowledge that the cell is nothing but the systematically organized community of molecular populations in dynamic interactions. And solely by learning more and more about the isolated pieces we can never hope to gain an understanding of the order to which the pieces are subordinated in those collective groupings which we know in cells ... The real cell is a physical continuum, no part of which can be considered as truly separate from the rest".

For these reasons he also criticized the habit of calling chemical compounds as "regulators", "integrators", "organizers" etc. without reference to the system organization in which they exhibit these properties and wrote [4]: "To state it bluntly, it would be rather a reversion of the prescientific age if on observing, for instance, the spinning of a whirl of fluid, one were to invoke a special compound as "spinner".

His mould breaking suggestions have, however, not been appreciated, possibly because he was himself not aware of the role of water as the material basis for the investigation of the unifying and differentiating aspects evident in the organism.

2. Water and the Unity of the Body

Water is contained in all parts of the body (Table 13 on p.77) and penetrates virtually all areas. It appears optimally "anchored" in all discontinuities of the body and to act as a mediator for all of the cycles between continuous and discontinuous regions. It has therefore been concluded that *water is a "conditio sine qua non"* for all molecular interactions within the body. Water provides the material basis for the unity and at the same time for the differentiation of the organism [9, 10].

Whereas structural changes in biomolecules can be determined in great detail, it is impossible to determine the differences in water structures with the same accuracy. Molecular biology is based on observable and measurable changes in the biomolecules [1], but the characteristic changes of water in all of the processes in a living organism cannot be investigated in this way.

This is because water has an enormous "potentia"[2] to integrate other forms in its oscillating network, whereby the integrated forms are considerably modified, so that water itself retains its main characteristics. In other words: information of other forms is integrated within the differentiation and the oscillating properties of liquid water in such ways that the changes of the water structure cannot be derived from the results of spectroscopic studies.

Liquid water has, therefore, the *unique ability to integrate other structures within its own oscillating structural network.* This requires that water must be able

(i) to "recognize" and "understand" information from other forms and

(ii) to "assimilate" and integrate them.

It may be said that water seems to "understand the languages of all other forms" and to "translate" all incoming information into its own language, the *water language.*

Water seems to be capable of "carrying" all information of the body. As in a hologram, the information is distributed over the whole system with each piece of information present in all of its parts. In water all of the information is stored and encoded in the *water language, which seems to be the universal language of the body, understood and used by all of its parts.*

Unfortunately, there is no direct access for man to understand this language and the information contained in water, a problem to which we shall draw our attention in the following section.

---

[1] The danger of exclusive considerations of structural aspects has been illustrated by *Weiss* [3] by comparing this with the inspection of a chemist's store with chemicals on shelves confined in bottles. The mere presence of them on the shelves a such has no effect whatever. However, when by design and choice some of them are opened and their contents mixed, predictable changes occur *as the outcome of organized behaviour.* Homogenization, by contrast, is comparable to the smashing of all bottles and mixing the contents of them all.

[2] *Potentia* manifests itself only in the *actus* and it is therefore outside the scope of a full mathematical, structural or thermodynamic characterization.

3. Water and Information

The word "information" is of Latin origin. "Informare" means to shape a form anew by adding and "working in" something from outside. Despite the lack of a clear definition of information, it is generally agreed, that information must serve a certain purpose and must be understood. The value and the understanding of information requires a "sender-receiver relationship". Both, sender and receiver must belong to the same system, in which the same language is spoken and understood. For example, during World War II the US-army transmitted messages across the Atlantic Ocean translated into the *Navajo* language, which was known only to the few members of the *Navajo* tribe. A few men of them on either side of the ocean acted as sender and receiver respectively.

With regard to the question of distinguishing between a (physico-chemically described) interaction and the exchange of information, we come to the following conclusion [9]: the concept of order differs from ordered arrangement in that the former requires finality (p. 175) and the latter only causality. By analogy, information is interaction in its final context whereas interactions as such may be described with the help of the laws of causality. In other words: *an interaction conveys information, without being information itself.*

The chief characteristics of information have been summarized as follows [9]:

(i) Transmission of information requires a certain basic conformity of sender and receiver, both of which must be parts of a superordinated system.

(ii) Information must have meaning and indivisibility; it is invariant towards changes in the molecular systems, and it cannot be localized within individual molecules.

(iii) This invariance is due to the fact that the molecules serve the whole system according to its requirements.

(iv) Information is maintained dynamically.

(v) Information must lead to a certain reaction in the receiver, i.e. the latter must be able to recognize, understand and use it.

(vi) Transmission of information is bound to lead to changes within the systems involved.

However, modern information theory is not concerned with finality or with understanding of information. Information processing is technically achieved as long as both sender and receiver are human beings who share the same understanding of the message transmitted. What is left, therefore, is the technical problem of transmitting the message in the form of a certain sequence of signals in an accurate manner.

According to *Shannon's* information theory [11] information is treated like a quantity that can be divided. In this way actually meaningless quantities are transmitted, as expressed for a computer as "garbage-in, garbage-out". It is up to the users of this equipment to formulate the given information in the form of a certain sequence of actually meaningless signals to allow the treatment of information as a stochastic process, one of the most common and efficient methods of probability calculus.

This situation may have contributed to a kind of belief in statistical results [3]. *It is frequently argued that liquid water cannot maintain information* because this seems unlikely, or even impossible, in view of the diffusionally averaged structure of liquid water with relaxation times of 10^{-10} seconds. This statement is, however, based on the result of a statistical interpretation of data, which concerns the quantitative characterization rather than the qualitative aspects of the actual structural and oscillating features and their information content. In the course of statistical evaluation all local and temporal differences within the liquid have been artificially eliminated.

The same procedure is applied in representing the crystal structure of a real crystal by means of the lattice parameters of the idealized model. This model illustrates very well the lattice-geometrical aspects in terms of equal quantities for unequal, but similar unit cells and hence it does not refer to the quality or to the information content of the real crystal under consideration: an ideal crystal would be devoid of any information content (see p. 175).

By analogy, the idealization of the diffusionally averaged structure of the liquid system under consideration cannot account for its system organization, for its quality and for its information content (see footnote 1 on p. 207). This situation provides therefore no argument against the information content of liquid water, but rather a challenge for its investigation.

In the course of the investigation we are, however, faced with an enormous problem, namely the above mentioned difficulty of gaining insight into the actual differentiation within the structural network and into the oscillating properties with all of the dissipative aspects which are bound to occur as new information is acquired or as integrated information is lost [4].

This problem is connected with the *limitations of man to understand the water language,* which, however, must be well understood by all of the interacting systems in all living organisms, although it is not directly accessible to the conscience of man.

4. Water and the Differentiation of the Human Body

Cell Membranes as Parts of the Water System

Faced with the high differentiation of the human body, the question may be raised, as to what ways the water system of the body provides not only for its unity but also for its differentiation.

The high differentiation on the cellular level is provided by the different properties and functions of intracellular and extracellular water. These are separated and interconnected by means of the cell membranes which consist of *bilayers* of

---

[3]   We are reminded of *Baierlein's* statement [12]: "If we want a theory of probability, we would do well to associate probability with degree of rational belief".

[4]   *Eigen* and *Winkler* [13] have pointed out that dissipative processes control and synchronize the information that is contained in the conservative elements of structure.

hydrophobic lipids, namely phospholipids, cholesterols and various proteins incorporated in them (p. 92).

The cell membranes are highly sensible to changes in water structures on either side of them and they show high permeability for water.

The cell membrane may be considered as the (meaningful) combination of two monolayers, one of them produced from intracellular water and the other one produced from extracellular water. Each of the monolayers is "anchored" by means of the hydrophilic head groups of the lipid molecules in its respective water system. This explains why each membrane has to different faces. The monolayers are held together by the interactions between the hydrophobic tails of the respective lipid molecules.

The internal monolayer provides the boundaries, i.e. the highest hierarchic level of intracellular water and the external monolayer provides the boundaries, i.e. the highest hierarchic level of extracellular water. The dissipative segregation processes and the mutual interactions of the highest hierarchic levels lead to formation of the bilayer with remarkably well developed static boundary conditions and at the same time with enormous fluidity and flexibility (p. 94). Its well developed conservative boundary conditions are dynamically maintained by the mutual interactions between the different aqueous subsystems by means of their respective highest hierarchic levels.

Intracellular water is stabilized by its interaction with the cell nucleus that contains the essential information of the cell. Extracellular water provides the cell with information from outside, i.e. from the organism, to which its inner part responds appropriately.

All changes near the hydrophilic groups are reflected in characteristic changes of the hydrophobic groups. For example, by the accumulation of ions near the hydrophilic groups water is consumed by the ions for hydration with partial dehydration of the hydrophilic head groups. These are, therefore, tightened with subsequent tightening of the hydrophobic tails.

The gradients between hydrophilic and hydrophobic groups have consequences for the transport of water molecules and of solutes within the membrane (p. 94). The solubility of water in hydrophobic media is by one or two orders of magnitude greater than the solubility of the hydrophobic substances in water (Table 14 on p. 97).

It has also been mentioned that water in hydrophobic capillaries is better organized than bulk water [5] and such capillary conditions are provided within the membranes. These are elastic, pipe-like channels, through which water molecules may be "blown through" with subsequent oscillations of the terminal tails. The specific properties of the capillaries are influenced by the environment of the hydrophilic groups (p. 93).

The hydrophobic groups of each monolayer seem, therefore, be informed about its water system and they appear to act on those of the other monolayer as a kind of

---

[5] This explains the role of the hydrophobic environment for the increased temperature range of supercooled water (p. 59 f).

highly sensitive sensor. In their flexibilities they are capable of "collecting" and "transmitting" information from either phase.

The functions of the protein groups are not fully understood, but they seem also related to the internal and external water systems respectively. It is known that the proteins in the outer layer are different from those in the inner layer.

On the other hand, the flexible hydrophobic layers are essential also for the efficient separation of extracellular and intracellular water. The hydrophobic double layer functions as a "barrier" between the two different liquid systems and in this way it is essential for the differentiation of the water system inside and outside the cell.

This barrier-function is reflected in the high electrical resistance of the membranes. The hydrophobic barriers help to maintain a certain "distance" between the aqueous phases and to direct their activities "inwards" so as to maintain their integral configuration and functionality.

The specific actions of each of the individual cells require different hydrophobic structures and properties in different cells. The membrane pattern of liver cells is different from that of skin cells. The conservation of the inner structure is greatest in the cells of the brain, which have a high water content and at the same time highly developed hydrophobic regions, the vibrations of which seem to be strongly influenced by their mutual interactions.

As the outer layer of the cell membrane is the highest hierarchic level of extracellular water and the inner layer the highest hierarchic level of intracellular water, they must represent the differences of their aqueous phases, respect each other and defend their functions according to the requirements of the whole organism.

The bilayer may, therefore, be considered as resulting from the dissipative segregation processes of intracellular and extracellular water with formation of highly developed conservative boundary conditions with the help of the lipid and protein molecules in the highest hierarchic levels of the two different aqueous solutions.

The membrane provides excellent conservative boundary conditions as well as enormous flexibility and adaptability and hence optimal conditions for the separation of intracellular and extracellular water by means of the flexible hydrophobic layers. As part of the aqueous system of the organism it provides at the same time all conditions for communication, exchange of information, matter and energy according to the requirements of the whole body and hence for its unity.

This state of optimal system organization of the membrane is characterized by highly developed and well-balanced conservative and dissipative features. This state of optimal system organization is further supported by the presence of water in the hydrophobic capillaries of the membrane and by the optimal system organization of water in the organism at the body temperature of + 37.5°C. The cell membranes provide, therefore, optimally organized regions within the water system of the organism.

The different roles of extracellular and intracellular water may be characterized as follows:

Intracellular water provides for the individuality of the cells and for the differentiation of the organism and hence the internal monolayer (as the highest hierarchic level of intracellular water) is highly differentiated.

Extracellular water is structurally less active and provides for the unity of the organism and hence its monolayer (as the highest hierarchic level of extracellular water) is less differentiated. It is known, for instance, that

(i) the proteins at the outer layer are different from those at the inner layer,

(ii) both the outer and the inner layer are influenced by changes in local composition near the hydrophilic groups, and

(iii) the hydrophobic groups, and the channels between them, are subject to changes which appear to be directed by the respective aqueous systems and their mutual interactions.

All of the actions of the cell membrane are therefore under the decisive influence of the differences of extracellular and intracellular water.

Boundaries Within the Cell

Each cell is part of the water system of the body and serves the purpose of the living body. If the cell is taken out of the body and grown in culture, it loses its finality and consequently its distinctive structure and individual properties.

Membranes are found within the cells, which provide adequate boundaries for functional subunits, namely the mitochondria, the microtubules and the microfilaments.

They are subordinated to the cell within which they have a certain autonomy. The *inner membrane of a mitochondrion* is the site of the individual respiratory functions that make the mitochondrion the so-called "power house" of the cell, that control the inward and outward passage of ions such as calcium ions and that regulate expansions and contractions of the mitochondrial surfaces according to decreasing and increasing respiratory activities. The *outer membrane* controls the passage of material into and out of the mitochondrion.

Genetic Information

The mitochondria contain hydrated DNA in the form of a "chromosome" which is surrounded by a dynamically active water layer. The enormous role of this water layer as required for the genetic information has not been studied in detail, but its non-freezeability down to temperatures of -60°C [14], the lowest temperature ever observed for "liquid water" (Fig. 61), is most exciting.

This behaviour shows that under the influence of the highly flexible DNA molecules, water is even better organized than in the supercooled state. The water system in the cell is characterized by the presence of various hydrophobic substances. The hydrophilic DNA molecules provide a kind of conservative boundary conditions for an excellent system organization which seems to be optimally developed at the border line area of water and DNA molecules.

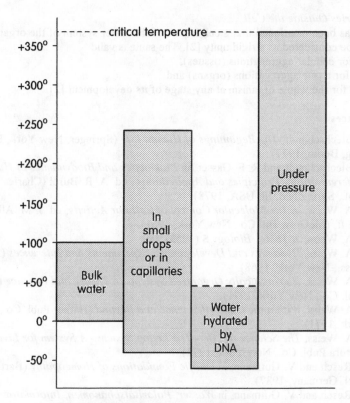

Fig. 61. Liquid ranges of water under different conditions.

With regard to the formation of these conservative boundary conditions, it should be borne in mind that - just as the cell membranes - *the DNA molecules are produced in highly specific ways within the water system of the organism.*

This requires that the informations necessary for the DNA formation must be provided by the water system within the cell as encoded in the water language. In these ways the highly developed conservative boundary conditions of the DNA structure are produced in vivo within the water system and dynamically maintained only within the union of DNA structure and its surrounding water system by the mutual interactions between them. When this union is destroyed by complete dehydration, the DNA structure is known to break down as well.

Whereas the role of water for the development of secondary, tertiary and quaternary structures has been studied [15], formation, reproduction and functions of DNA structures have - so far - not been considered as governed by the dissipative actions of the highly organized water system of the body.

Boundaries Outside the Cell

It has been mentioned that a cell is part of the water system of the organism and cannot be considered as a rigid unity [2]. The same is valid
(i) for cellular aggregations (tissues),
(ii) for tissue aggregations (organs) and
(iii) for the whole organism at any stage of its development [2].

References

1. E. Blechschmidt, *The Beginnings of Human Life* (Springer, New York, Heidelberg, Berlin, 1977).
2. E. Blechschmidt and R. F. Gasser, in *Biokinetics and Biodynamics of Human Differentiation - Principles and Applications* , ed. A. R. Burdi (Charles Thomas Publ., Springfield, Ill. USA, 1978).
3. P. A. Weiss, in *The Molecular Control of Cellular Activity*, ed. J. M. Allen, p. 1 ff, (McGraw Hill Co., New York, 1961).
4. P. A. Weiss, *J. Theor. Biology* 5 (1963) 389.
5. P. A. Weiss, *Dynamics and Development: Experiments and Inferences* (Acad. Press., New York, 1968).
6. P. A. Weiss, *Hierarchically Organized Systems in Theory and Practice* (Hafner Publ. Co., New York, 1971).
7. P. A. Weiss, *Within the Gates of Science and Beyond* (Hafner Publ. Co., New York, 1971).
8. P. A. Weiss, *The Science of Life - The Living System - A System for Living* (Futura Publ. Co., New York, 1973).
9. G. Resch and V. Gutmann, *Scientific Foundations of Homeopathy* (Barthel Publ. Germany, 1987).
10. G. Resch and V. Gutmann, in *Wasser, Polaritätsphänomen, Informationsträger, Lebens- Heilmittel,* ed. I. Engler, p. 193 ff, (Sommer Verlag, Teningen, Germany, 1989).
11. C. E. Shannon in *Bell. Syst. Technical Journal* 27 (1948) 3, 379, 623.
12. R. Baierlein, *Atoms and Information Theory*, 2nd ed. (Acad. Press., New York, 1962).
13. M. Eigen and R. Winkler, *Das Spiel,* (Piper Verlag, München, Zürich, 1984).
14. M. Falk, A. G. Poole and C. G. Goymour, *Can. J. Chem.* 48 (1970) 1536.
15. D. Eagland, in *Water, a Comprehensive Treatise,* ed. F. Franks, Vol. 4, p. 305, (Plenum Press, New York, 1975).

CHAPTER 20

ORGANIZATION IN NON-AQUEOUS SOLUTIONS

1. Protic Solvents and their Solutions

The fundamental elements of system organization in liquid solutions have been derived for liquid water and its solutions by means of studies under the most complex conditions. Closest resemblance to water is found for liquids which are known as "hydrogen-bonded" or "structured" liquids such as liquid hydrogen fluoride, liquid ammonia, anhydrous acetic acid, sulphuric acid and monohydric alcohols [1].

All of these liquids may be characterized by
(i) hydrogen-bonded solvent molecules (p. 73),
(ii) the presence of a self-ionization equilibrium (Table 8 on p. 73),
(iii) the presence of dissolved gases (Table 16 on p. 98),
(iv) the decisive role of the interface.

Their system organization may therefore be illustrated in an analogous manner as that for liquid water, when the terms "hydrophobic" and "hydrophilic" are replaced by "solvophobic" and "solvophilic" respectively. In this way the most general illustration for the system organization of a liquid is represented in Fig. 62.

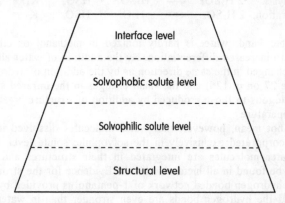

Fig. 62. Illustration of the system organization in liquid solutions.

The freezing points of anhydrous hydrogen fluoride and ammonia, which are much lower than those of water, indicate less developed structural features. Whereas in liquid water a three-dimensional network of great strength and high flexibility is present, one-dimensional molecular chains are found in liquid hydrogen fluoride and in liquid ammonia. In the monohydric alcohols the structural network is looser than that in water because of the presence of the hydrophobic alkyl group in each alcohol molecule.

These structural differences are related to the more restricted solubilities and reactivities of non-aqueous solvents. They indicate less developed ability for adaption and for integration and hence a less developed system organization in the said liquids.

Another point of interest is the fact that none of these solvents can be completely freed from the last traces of water. This leads one to the question if the presence of water might be of importance for the very existence of the liquid and its system organization.

The high mutual solubilities in protic solvents and water - in many cases complete miscibilities - indicate that water can integrate their structures and that water can be integrated in the liquid structures of the protic solvents.

This integration may be accomplished by the well-developed ionizing properties of these solvents, by which water is converted into H_3O^+ ions in acidic solvents such as hydrogen fluoride, acetic acid, or sulphuric acid and into OH^- ions in basic solvents such as liquid ammonia (Table 8 on p.73). In these ways water contributes to increased differentiation of the solvophilic solute level (see Fig. 61), i.e. the level that is immediately superordinated to the lowest hierarchic level.

In "anhydrous" sulphuric acid this differentiation is further accomplished by the production of H_3O^+ ions due to its *self-dehydration* equilibrium, so that sulphuric acid of ideal composition H_2SO_4 cannot be obtained [2]:

Autoprotolysis: $2 H_2SO_4 \rightleftharpoons H_3SO_4^+ + HSO_4^-$; $K_1 = 2.10^{-4}$
Self-dehydration: $2 H_2SO_4 \rightleftharpoons H_3O^+ + HSO_4^-$; $K_2 = 2.10^{-3}$

On the other hand, water is hardly ionized in methanol or ethanol, but its presence leads to increased differentiation. By the addition of water all macroscopic properties are changed in the same direction as by the addition of hydrophilic solutes to water (Table 27 on p. 126) and the small changes in the infrared spectrographic pattern are analogous to those produced by addition of structure breakers [3] or by decrease in temperature.

This does not mean, however, that water molecules dissolved in monohydric alcohols are incorporated exclusively in the solvophilic solute level. In fact, small amounts of water molecules are integrated in their structures and hence water molecules may be found in all hierarchic levels. Evidence for the strong integration of water in the hydrogen bonded network of 1-pentanol is provided by *Rosenholm*'s finding [4], that the hydrogen bonds are even stronger than in water. *Bonner* [5] concluded from spectral results in the overtone range that "water appears to break up the alcohol structures" and this implies a great structurizing influence of water on alcohol.

Optimal integration of water in ethanol (complete "penetration") seems to occur in a solution of 4.43 wt-% water in ethanol (azeotropic mixture boiling at +78.15°C) and at low temperature in a solution of 7.6 wt-% water in ethanol (eutectic mixture freezing at -123°C) (Table 22 on p.119).

The liquid range of the lower aliphatic alcohols extends to much lower temperatures than that of water (Table 19 on p.117). This indicates that in the liquid alcohols the conservative boundary conditions are less developed than in water, i.e. their network is looser and more reluctant to integrate matter, energy, and information.

2. Aprotic Solvents and their Solutions

"Structured" Solvents

Aprotic solvents are frequently considered as "non-structured" and this applies to the lack of structural information of all of the organic aprotic solvents. There are, however, a few inorganic aprotic solvents for which a certain structural characterization is at least indicated.

The extension of the *Brönsted*-concept of the proton transfer between acids and bases to halide ion transfer reactions [6] in solutions of bromine(III)-fluoride, iodine monochloride or arsenic(III)-chloride lead to the suggestion of the presence of halide ion bridges in these solutions [1,7,8].

The chain-like structure of iodine monochloride is similar to that of liquid hydrogen fluoride and there are certain similarities in structural features in liquid ammonia and in liquid bromine(III)-fluoride:

In further analogy to the "structured" protic solvents, self-ionization equilibria are established in all of the said solvents [7,8] and hence solvent ions are always bound to be present in the purified solvents. This means that a certain differentiation of the solvophilic solute level is always provided in these liquids. The differentiation is further increased by the unavoidable presence of water in all of these liquids, which is readily hydrolysed by them. In this way the solvophilic level is further differentiated.

The role of the gases dissolved in these solvents has not been studied in detail, but their unavoidable presence cannot be questioned.

It may therefore be suggested that the elements of the system organization of aprotic "structured" liquids is analogous to that found in hydrogen bonded solvents as presented in Fig. 62.

Other Aprotic Solvents

The situation is somewhat different with regard to the great number of solvents and their solutions where no experimental evidence is available about their liquid structures. For this reason, nothing is known about the influence of the changes in the higher levels on the changes in the lowest level. Efforts in the chemistry of aprotic solutions are concentrated on the influence of the bulk solvent on the solute structures and properties, the solvent properties being characterized on the molecular level by empirical solvent parameters (Chapter 13) [9].

The absence of knowledge on structural features is paralleled by the absence of self-ionization equilibria. This means that the differentiation in the hydrophilic level must be due to the unavoidable presence of other solutes.

One of those solutes which cannot be completely removed is water. In all of these solvents water is present in minimum concentrations of 10^{-6} mol/l (p. 131) and this concentration is somewhat greater than that of the solvent ions produced by self-ionization in liquid water.

With regard to the role of water in these solutions a distinction may be made between

(i) solvents which dissolve water readily (hydrophilic solvents) and

(ii) solvents which show extremely low solubilities for water (hydrophobic solvents).

In an aprotic hydrophilic solvent such as acetone or acetonitrile, donor-acceptor interactions with water readily take place [9]. These lead to local structural contractions and hence to an improvement of the conservative boundary conditions. This means that water is integrated mainly in the solvophilic solute level (Fig. 63).

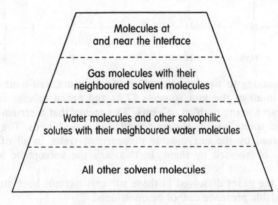

Fig. 63. Illustration of the system organization of a hydrophilic aprotic solvent such as acetone or acetonitrile.

The role of water may be even more decisive for the existence of a hydrophobic liquid such as benzene or carbon tetrachloride. The low solubilities of water in these solvents (Table 14 on p.97) are comparable to gas solubilities in water. It has been suggested that the water molecules in these solvents might be entrapped in holes of the liquid structure which may be stabilized in a manner analogous to the way holes are stabilized in water by the presence of entrapped gas molecules. The thermal stabilities of solid clathrates are known to be increased by the presence of "help gases" (p. 100). As the molecules in aprotic solvents seem to have more "gas-like" character than those in structured solvents, it may be suggested that the water molecules may help to improve the static boundary conditions of the liquid structure by increase of the stabilities of the inner surfaces. In this way water molecules would have a high hierarchic significance by serving on the level that is immediately subordinated to the highest level (Fig. 64) and contribute to the stabilization of the liquid state.

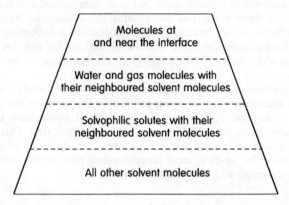

Fig. 64. Illustration of the system organization of a hydrophobic aprotic liquid, such as benzene or carbon tetrachloride.

The system organizations in aprotic solvents - as compared to that of water - are not highly developed, apparently because of the less pronounced conservative boundary conditions, but the stabilities of the liquids may be due to the role of water in the hydrophobic solute level that is immediately subordinated to the highest hierarchic level (Fig. 64).

3. Changes in Organization

Influence of Solutes

Amphipathic solutes are known to have dramatic influences on the liquid properties even when they are present in very low concentrations. This is because they tend to occupy positions in the highest hierarchic level with substantial lowering of the surface energy and with dominating influence on the lower levels.

Solvophobic solutes take positions in the level that is immediately subordinated to the highest level. Since the number of positions in this level is greater than that in the highest hierarchic level and since the influence of this level on the liquid properties is less pronounced than that of the highest hierarchic level, a somewhat higher concentration of hydrophobic solutes is needed in order to produce substantial changes in the properties of the liquid.

Even less pronounced is the influence of small amounts of solutes which occupy positions in the solvophilic level. In solvent mixtures the solvent molecules of better developed coordinating properties are known to have a greater share in solvation phenomena and this is described by preferential solvation (p. 152 ff) [10].

The smallest changes in properties of a solution are observed by addition of solutes which may be well-integrated in the lowest hierarchic level. For example, the properties of water are hardly changed by addition of small amounts of ethanol. As its concentration is increased, however, the dynamic properties are improved as is seen from the decrease in freezing point and in surface tension and from increase in heat capacity. As these properties are considerably influenced by integration of solutes in the higher levels, ethanol molecules lead to a loosening of the water structure by taking positions in the solvophilic solute level [8] and this has also been concluded from the results of IR-spectrographic measurements on water-ethanol mixtures [11].

Alcohol molecules may even be found in the solvophobic level of water, as some of the ethanol molecules may be entrapped within the voids of the solution [12]. Structural work has revealed that each of these ethanol molecules is surrounded by an "inner-surface" made up of 17 water molecules (clathrate). There are several indications for a well-developed system organization in the water-ethanol system in the range of about 50 wt-% ethanol corresponding to a molar ratio of water to ethanol of approximately and 3:1 (see p. 121 ff).

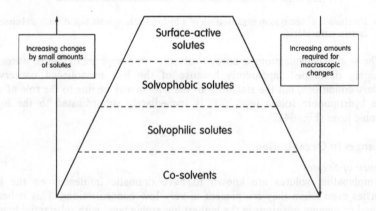

Fig. 65. Illustration of the effects of solutes for changes in solution properties.

Fig. 65 illustrates the fact that *low concentrations of a given solute change the solvent properties to a greater extent, the higher the hierarchic level in which the solute is mainly integrated* [13].

Changes in Macroscopic Properties and in System Organization

It has been emphasized that changes in system organization must be indirectly inferred from all observable (and measurable) changes under complex conditions. It may therefore be of interest to learn in what ways changes in macroscopic properties may be related to changes within the system organization of a given liquid.

Because each solution acts as a unity, it is, unfortunately, impossible to attribute one of the changes of a certain macroscopic property to a change in a certain hierarchic level. In addition, the quantitative results for the characterization of macroscopic properties are usually obtained by the application of statistical methods. In order to obtain certain indications for changes in the various hierarchic levels, the changes of different properties must be considered, but reliable conclusions cannot be drawn.

An increase in *entropy* is always due to an increase in system differentiation and in increased dynamic properties of the system, whereas decrease in entropy is an indication of increasing conservative structural aspects.

The changes in *vapour pressure* and in *surface tension* are expressions of changes mainly occurring in the *highest hierarchic level*. Increase in dissipative actions is indicated by increase in vapour pressure and by decrease in surface tension.

Increase in *specific heat* is related to increase in dynamic features either by increase in temperature or by increase in gas content and hence to an increase in dynamic properties in the *hydrophobic level*.

Changes in *electrical conductivity* and *thermal conductivity* are due to molecular or energetic changes in the *lower levels*. Such changes are also indicated by increase in *viscosity* and in *density*, which are usually associated with an increase in conservative boundary conditions and these are also strengthened by decrease in temperature.

Table 36. Indications for assigning changes in macroscopic properties to changes in different levels.

Changes in the interface level are indicated by changes in:	Changes in the hydro-phobic level are indicated by changes in:	Changes in the lower levels (hydrophilic and structural) are indicated by changes in:
surface tension vapour pressure	gas solubility specific heat	conductivity density viscosity spectroscopic and structural data

These considerations are in agreement with the opposite effects observed by addition of solvophobic and of solvophilic solutes to water (Table 18 on p.104).

Table 37 may serve for a first orientation for changes in conservative and dissipative aspects.

Table 37. Changes in conservative and dissipative aspects as related to changes in certain macroscopic quantities.

Increasing values in	Conservative Aspects	Dissipative Aspects
Entropy	decreasing	increasing
Specific Heat	decreasing	increasing
Surface Tension	increasing	decreasing
Vapour Pressure	decreasing	increasing
Compressibility	decreasing	increasing
Density	increasing	decreasing
Viscosity	increasing	decreasing
Thermal Conductivity	increasing	decreasing
Electrical Conductivity	increasing	decreasing

Remarks on Homogeneous Catalysis

A catalyst must be highly organized with sufficient static boundary conditions (which are dynamically maintained) while acting dynamically on its environment without losing its integral configuration and dominating influence.

It has been mentioned on p. 202 that states of optimal system organization are obviously involved in the course of heterogeneous catalysis.

The optimal organization of the whole system may be due to the establishment of a "balance" between well-developed conservative boundary conditions of the solid and the highly developed dissipative actions in the gas phase, which are directed in a certain way under the control of the former.

With regard to homogeneous catalysis in solution both catalyst and substrate may act on the same hierarchic level, usually the solvophilic solute level of the liquid system. The conservative and dissipative aspects in the liquid are therefore more "balanced" than either in the solid or in the gaseous state.

This "balance" appears to be "improved" in the solution by addition of the appropriate catalyst, which, by definition, has the ability to influence its environment while maintaining its chief characteristics and this is characteristic of a highly organized system.

The highly developed organization of the whole system requires appropriate organization forms not only on the supermolecular but also on the intramolecular level of the interacting molecules. Analogous considerations apply for an understanding of all system considerations. For example, the observations made in the course of the reduction of tris(phenantroline)iron(III) in dimethylsulfoxide namely the nearly temperature independent reaction rate [14], imply that the energetic changes due to changes in temperature must be redistributed over the whole solution in such ways that the solute species remain nearly unaffected (p. 150) [15]. The solution seems to be in a position to provide energy to the solutes on cooling and to integrate the gain in energy on heating in such ways that the reaction rate remains unaffected.

Any supermolecular system organization implies that the solute molecules must be organized within themselves and this leads to the consideration of *intramolecular system organization*.

References

1. V. Gutmann, *Coordination Chemistry in Non-Aqueous Solutions* (Springer, Wien, 1968).
2. R. J. Gillespie, *J. Chem. Soc.* (1950) 2516.
3. C. Kuttenberg, E. Scheiber and V. Gutmann, *Monatsh. Chem.* in the press.
4. J. B. Rosenholm, *Ber. Bunsenges. Phys. Chem.* **91** (1975) 106.
5. O. D. Bonner and Y. S. Choi, *J. Sol. Chem.* **4** (1975) 457.
6. V. Gutmann and I. Lindqvist, *Z. Phys. Chem.* **203** (1954) 250.
7. V. Gutmann, *Quart. Revs.* **10** (1956) 451.
8. V. Gutmann, *J. Phys. Chem.* **63** (1959) 378.
9. V. Gutmann, *The Donor - Acceptor Approach to Molecular Interactions* (Plenum, New York, 1977).
10. W. J. MacKellar and D. B. Rorabacher, *J. Amer. Chem. Soc.* **93** (1971) 4379.
11. V. Gutmann, G. Resch, G. Scheiber and C. Kuttenberg, unpublished.
12. V. Gutmann, *Pure Appl. Chem.* **63** (1991) 1715.
13. V. Gutmann and G. Resch, *Atti. Accad. Peloritana, Classe I* **70** (1992) 59.
14. R. Schmid, R. W. Soukup, M. K. Aresteh and V. Gutmann, *Inorg. Chim. Acta* **73** (1983) 21.
15. V. Gutmann and G. Resch, *Monatsh. Chem.* **116** (1985) 1107.

CHAPTER 21

INTRAMOLECULAR SYSTEM ORGANIZATIONS

1. Tris(phenantroline)-iron Complexes

Primas [1] pointed out that the *simplest hierarchic system is a molecule*, but he made no attempt to investigate the intramolecular system organization. The first report on *intramolecular system organization* has been published for tris(phenantroline)-iron complexes [2]. The unusual experimental results (p. 147 ff) [3] lead to the conclusion that the electron densities at the coordination centre remain nearly invariant

    (i) to changes in its redox state (p. 147),

    (ii) to changes caused by substitutions at various ligand positions (p. 148) and

    (iii) to changes of the solvent.

By considering further

    (i) the solvent effects on the rate coefficients of redox reactions [4],

    (ii) the isokinetic relationships and

    (iii) the unusual polarographic results [5],

the conclusion has been reached that these complexes have an unusually well-developed ability of adapting themselves to different environmental conditions without losing their integral configuration [2,6,7].

By means of similarity considerations the differentiation within the structural pattern of the complex ions and the different resistance towards various forces have been worked out.

The iron atom is readily distinguished from all other atoms. With regard to the six nitrogen atoms, we are confronted with the problem of their multiplicity. Their individual properties cannot be derived from the results of spectral analysis, but the results of quantum chemical calculations [3] reveal slight differences (Fig. 45).

Although according to the idealized structural representation in Fig. 45 the hydrogen nuclei may be expected to have the same net charges, they are subject to continuous slight changes because of the movements of the phenantroline groups within the changing conditions of the solution. Differences are also produced by the presence of water molecules within at least one of the pockets formed between the coordinated phenantroline groups. The quantum chemical calculations provide different values for each of the nitrogens and for each of the terminal hydrogen atoms (Fig. 45 on p. 148).

As molecules react specifically towards their environment, the differences between the parts must be intentional with respect to the overall situation. This requires certain local and temporal differences in energy and hence superordinations and subordinations.

The greatest invariance towards different conditions is found for the charge density region within the FeN_6 group. This group appears, because nearly unaffected by electron changes itself, to redistribute the charge density within the whole system. This group may be structurally considered as the inner region of the pockets provided by phenantroline ligands and these are particularly active in interactions with the environment. Water molecules within the pockets may act as bridging units between the molecular centre and the molecular environment.

All of the said properties suggest that each FeN_6 group acts as a regulating centre, a kind of "turn over place" for the redistribution of electron densities with appropriate changes in nuclear positions within the structural framework of the complex unit. This group is therefore considered as the highest accessible hierarchic level.

Immediately subordinated to this level appear the boundary areas of the ligands which are essential for the communication with the environment [7] (Fig. 66).

The π-electron systems of the aromatic rings show greater mobility than the other electrons within the structural framework. The π-electrons must be capable of executing either donor or acceptor functions towards other areas [3] and hence they appear to play a decisive role by exercising "buffer functions" between the higher levels and the structural framework. It is therefore appropriate to consider the π-electron system as serving on the level which is between that of the boundary areas and that of the molecular framework.

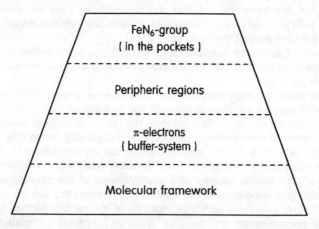

Fig. 66. Illustration of the intramolecular system organization of tris(phenantroline) iron complexes.

All other electron areas of the complex entity appear subordinated to this level. The more pronounced dynamic aspects of the superior levels are necessary in order to maintain essentially the structural framework (provided by the lowest hierarchic level), which provides the boundary conditions for the execution of the dynamic functions of the higher levels.

Whereas the structural boundary conditions provided by the lowest level are well accessible, it is impossible to account for all details of the intramolecular system organization, namely

(i) the differentiation within the electron areas of the FeN_6 group,

(ii) the differentiation and oscillations in the peripheric region and

(iii) the variations within the π-electron system [2,3].

It has been pointed out on p. 150 that the changes in magnetic properties in the course of redox reactions cannot be ascribed to localized changes at the coordination centre, but rather to result from the redistribution of electronic changes over all of the complex ion.

The question may also be raised as to whether there is an association between high invariance of the charge density around the coordination centre and the establishment of a high symmetry in charge density around the latter. For octahedral coordination by equal ligands one would expect a fairly high symmetry in the charge density distribution. The octahedral tris(phenantroline)-complexes are in low spin states. The low spin state is maintained only as long as the nitrogen atoms within the ligand molecules remain in similar environments.

Substitutions at 2,3,4,5,6 or 7-positions do not provoke substantial changes around the nitrogen atoms, but the situation is different for substitution in the immediate neighbourhood of only one of the nitrogen atoms, for example by substitution only in the 1-position or only in the 8-position. In these cases the Fe-N-distances no longer remain equal. The internal tensions cannot be overcome within the rather rigid framework of the low spin complex and the system has to gain a dynamically better developed state by transition from the low spin state into the high spin state [8].

In this state of higher energy the system is characterized by greater differentiation [9-11]. In the course of the low spin → high spin transition the system passes a state of *highest differentiation and greatest mobility*, in which the static and dynamic aspects of order appear to be "well-balanced", the system is *optimally organized*.

Because the complexes under consideration can reach this state at room temperature, it may be concluded that they are highly organized at room temperature.

2. Solvatochromic Complexes

Bis(phenantroline)iron(II)cyanide

It has been shown on p. 140 ff that bis(phenantroline)iron(II)cyanide shows solvatochromism and can be applied as a colour indicator for the estimation of the acceptor properties of non-aqueous solutions [13-18].

The spectral properties of this complex remain nearly invariant to substitutions at the phenantroline ligands. The *Mössbauer* isomer shift at the iron nucleus is similar to that in the tris(phenantroline) complexes and this indicates similar electron densities around the iron nucleus. The coordination centre is situated in the pockets between the ligands and hence it is an integral part of the outer regions. The charge density pattern around the coordination centre appears to retain locally a nearly invariant charge density, as the acceptor strength of the solvent is changed. This implies that all changes in electron densities due to coordination are redistributed over the whole species in highly specific ways.

This redistribution appears to be facilitated by the ease of electron shifts within the back-bonding regions at the cyanide groups and the π-electron systems of the phenantroline ligands.

In analogy to the tris(phenantroline)-iron complexes the region around the iron nucleus is considered as the highest hierarchic level by which the electron densities are regulated in all other regions, in particular within the cyano groups. In all other regions the charge density is subject to the regulating actions by this centre and hence they are subordinated to it.

The redistribution is effected by means of the highly mobile π-electron systems in cooporation with the back-bonding regions and hence the system organization may be illustrated in the same way as presented in Fig. 66.

Copper(II)-Complexes with Tetramethylethylenediamine and a β-diketone

As described on p. 138 f the above mentioned complexes provide useful colour indicators for the estimation of solvent donor properties [19]. The planar complexes, represented on p.140, undergo changes in colour as donor units approach the coordination centre with formation of a pseudo-octahedral geometry. The original and planar arranged ligands retain essentially their positions as equatorial sites within the developing pseudo-octahedral configuration. As the donor molecules approach the coordination centre, the Cu-O bonds and the C-C bonds are lengthened and the C=O bond is shortened.

The redox potentials of the first and of the second reduction step in different solvents are nearly linearly related to the solvent donor numbers and, except for acetonitrile, an isoequilibrium temperature is found at +8°C [20] (Fig. 67):

At the isoequilibrium temperature the redox equilibria under consideration are independent of the donor strength of the solvent and hence independent of the actual structural arrangement as well as of the energy state of the dissolved complex. Since redox behaviour concerns the whole species within its environment, the differences in energies and charge densities in different solvents must be distributed over the whole system in highly specific ways. This implies the establishment of a particularly well-developed system organization of each of the solutions at +8°C.

The supermolecular system organization of a liquid solution has been illustrated in a general way in Fig.62. The intramolecular system organization within the complex species may be similar to that as represented in Fig. 66.

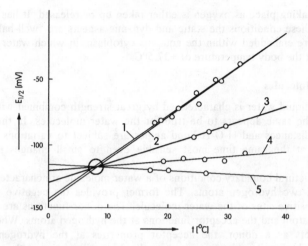

Fig. 67. Van't Hoff plot of the half-wave potentials of the second reduction step of
[Cu(tmen)(acac)]NO₃ vs. temperature. 1 Nitroethane, 2 Dimethylsulfoxide,
3 Dimethylformamide, 4 Dichloroethane, 5 Acetone.

3. Haemoglobin

The behaviour of haemoglobin has been explored in much detail by *Perutz*
[21,22]. It provides an example for the interdependence with the boundary
conditions within the erythrocyte and for the fact that it serves a highly specific
purpose, namely to carry oxygen from the lungs to the tissues and to provide the
conditions for the transport of carbon dioxide from the tissues to the lungs.

The structural and conformational changes have been worked out in great detail
[21,22] as briefly outlined on p. 35 ff. These provide indications for its system
organization but there are many points which demand further exploration.

For the coordination of oxygen four iron atoms are available, each of them
situated at a haeme-group in one of the pockets of the molecule. It is not understood
in which ways the flux of information between the iron atoms is taking place. As one
of them has taken up an oxygen molecule, the others will immediately follow suit
and the same is true for the release of oxygen at the tissues. These interactions are
also known as haeme-haeme interactions.

It has been outlined on p. 35 that the weak interactions of oxygen molecules at
the iron nuclei cause not only dramatic conformational changes, but also proton
transfer reactions at the ends of the chains with the cytoplasm.

The electron densities around the iron atoms seem to be part of the highest
hierarchic level of the molecule. Immediately subordinated appear the terminal
molecular regions which are in continuous relations to the cytoplasm.

The intramolecular system organization of haemoglobin within the erythrocytes
of the human body is highly developed. This is indicated by the spin-crossover

transitions taking place as oxygen is either taken up or released. It has been shown that under these conditions the static and dynamic aspects are "well-balanced". The molecules are embedded within the aqueous cytoplasm in which water is optimally organized at the body temperature of +37.5°C.

4. Water Molecules

Just as liquid water is characterized by great strength combined with enormous flexibility, the same appears to be true for the water molecules. In the molecules, O-H bond distances and H-O-H bond angles are subject to variations to a certain extent and at the same time most sensible even to small changes in molecular environment.

The structural boundary conditions of a water molecule are characterized by one oxygen and two hydrogen atoms. The former provides the negative site and the latter the positive sites of the water molecule. The donor functions are exercised at the oxygen atom and the acceptor functions at the hydrogen atoms. When the water molecule acts as a donor, the acceptor properties at the hydrogen atoms are increased and when it acts as an acceptor, the donor properties at the oxygen atom are increased.

Because the donor functions involve greater activities than the more passive acceptor functions and because the oxygen atom has a more central position in the molecular structure, it may be suggested that the charge densities at the oxygen atom have a higher significance in the system organization of the water molecule than the hydrogen atoms. This is also indicated by the stronger hydration of cations than that of anions of the same size and charge, and this shows that water molecules are more easily oriented with the (hierarchically higher) oxygen sites inward to the positively charged ions than with the (hierarchically lower) hydrogen sites inward to the negatively charged ions.

The actual "organizing centre" within the water molecule cannot be clearly localized within its molecular structure, but it may be expected to be under the immediate sphere of influence of the oxygen atom.

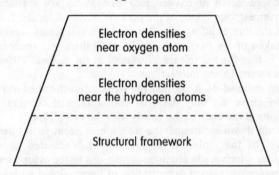

Fig. 68. Illustration of the intramolecular system organization of a water molecule.

We suggest, therefore, that one should consider the charge density pattern at (the inner site of) the oxygen atom as the highest hierarchic level that is accessible. To this the outer regions of the hydrogen atoms are subordinated. The lowest level appears to be provided by the electron densities between the atoms and these are mainly responsible for maintaining the molecular framework of the water molecule. This has been illustrated in Fig. 68.

Each of the water molecules has its "individuality" within the complex relationships. This has been pictorially illustrated like men with their highest hierarchic level represented by their head (for the water molecules the oxygen atoms) "walking" or "dancing" according to a certain melody [6]. In the cartoon below the water molecules have been assigned a high intelligence quotient [1] [6].

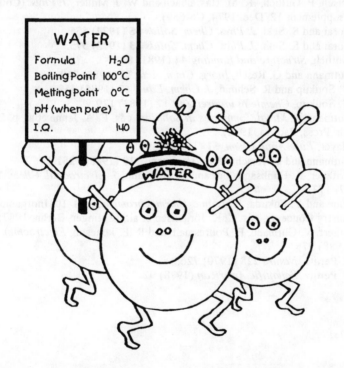

[1] By request of the authors [6] the following statement was published in due course: "The article submitted was originally entitled 'New frontiers of the molecular concept'. Prof. Gutmann comments, to change the title to 'Intelligent molecules' may help to make our ideas more popular, although this formulation could not be maintained from our philosophical point of view. The molecules follow an intelligent pattern, which is a prerequisite for their actions. In this sense, it is also not correct to talk of intelligent products. Since this term is widely used and accepted, I have no sharp rejection to this change of title, although I do not feel perfectly happy with it" (see also p. 37, including the footnote there).

References

1. H. Primas, *Chemistry, Quantum Mechanics and Reductionism,* (Springer, Berlin, Heidelberg, New York, 1983).
2. V. Gutmann and G. Resch, *Monatsh. Chem.* **116** (1985) 1107.
3. W. Linert, V. Gutmann, G. Wiesinger and P. G . Perkins, *Z. Physik. Chem.* (n.F.) **142** (1985) 221.
4. R. Schmid, R. W. Soukup, M. Aresteh and V. Gutmann, *Inorg. Chim. Acta* **73** (1983) 21.
5. S. A. Saji, T. Fukai and S. Aoyogui, *J. Electroanal. Chem.* **66** (1975) 81.
6. V. Gutmann and G. Resch, *Chem. Int.* **10** (1988) 5.
7. V. Gutmann and G. Resch, *Atti Accad. Peloritana, Classe I* **70** (1992) 59.
8. J. Fleisch, P. Gütlich, K. M. Hasselbach and W. J. Müller, *J. Phys.* (Collogue C 6, supplement 12, Dec. 1974, C6-659).
9. M. Sorai and S. Seki, *J. Phys. Chem. Solids* **35** (1974) 555.
10. M. Sorai and S. Seki, *J. Phys. Chem. Solids* **33** (1972) 575.
11. M. Gütlich, *Structure and Bonding* **44** (1981) 84.
12. V. Gutmann and G. Resch, *Inorg. Chim. Acta* **72** (1983) 269.
13. R. W. Soukup and R. Schmid, *J. Chem. Educ.* **62** (1985) 459.
14. R. W. Soukup, *Chemie in unserer Zeit* **17** (1983) 129.
15. V. Gutmann in, *Metal Complexes in Solution,* eds. E. A. Jenne et al, p. 205, (Piccin Press, Padova, 1986).
16. U. Mayer, *Pure Appl. Chem.* **51** (1979) 1697.
17. V. Gutmann and G. Resch, *Monatsh. Chem.* **119** (1988) 1251.
18. G. Gritzner, K. Danksagmüller and V. Gutmann, *J. Electroanal. Chem.* **72** (1967) 177.
19. K. Sone and Y. Fukuda, *Inorganic Thermochromism,* Vol. 10, Inorganic Chemistry Concepts, eds. C. K. Jörgensen et al., (Springer, Berlin, 1987).
20. W. Linert, V. Gutmann, B. Pouresmail, and R. F. Jameson, *Electrochim. Acta* **33** (1988) 975.
21. M. F. Perutz, *Nature* **228** (1970) 726.
22. M. F. Perutz, *Scientific American* (1978) 92.

EPILOGUE

Due to the unorthodox character of these lecture notes, the reader may be left with a number of difficulties and problems. He has been used to concede that theories may be modified and new concepts introduced only as long as they fit into the general framework of the theory of science, which frequently is considered as an article of faith and not subject to challenge. Consequently, empirical sciences are described in terms of theories, which - according to *Popper* are "nets cast to catch what we call the world: to rationalize, to explain, and to master it. We endeavour to make the mesh ever finer and finer".

It is, therefore, difficult to understand the demand made by the authors, not to confuse the concept of system organization with a new theory or a new paradigm, but rather to accept it as logical consequence of one of the principles in nature. Unlike a theory, a principle is not established by scientists, but rather found by philosophers by completing the process of induction. The principle, found by philosophers, must, however be respected by the scientist in the course of every investigation, because each science must be founded on a principle.

As a consequence of the principle of motion, the system organization is a "conditio sine qua non" for each material system. It is required for the very existence and the observability of each object or subject. It is related to quality, the expression of reality rather than of human imaginations about it.

The proposed approach requires, therefore, a change in attitude by the scientist towards nature and towards the way to acquire knowledge. He has to start from observations of the qualities and to use his intellect for *reading into the things as they are* and not into our own imaginations about them.

In the attempt to read into the things, the exclusion of metaphysics from science and philosophy cannot any longer be maintained. In fact, metaphysics has never been completely excluded from scientific investigations, since an abstraction involves metaphysical criteria.

It would have been desirable to deal with this fundamental problem in extenso. Unfortunately, this would have been outside the scope of lecture notes on solution chemistry. The authors of this book hope, however, to provide this presentation as a separate volume tentatively entitled "Lecture Notes on Philosophy of Scientific Cognition" in the near future.

The philosophical considerations show that every system that can be perceived must have a hierarchic structure; without such hierarchic structures a system in its differentiation cannot react as a unity. The investigation of these structures have led us to a new understanding of the liquid state, especially of liquid water and its solutions.

The essential role of water in all living organisms is obvious to everybody but has not yet been sufficiently investigated. However, these questions can only be examined by considering the potentialities of water as they are exhibited only in the complex relationships in natural systems.

In this way it should be possible to establish, among the enormous amount of results available from the various branches of science, the common features which would allow us to locate each individual item of information within the superordinated relationships.

Especially in biology and medicine the countless disconnected findings available could be used to understand the whole. For it is only through an understanding of the superordinated relationships that the generally valid aspects of partial results can be recognized and brought into a correct connection to other partial results.

The fate of this approach, however, cannot depend on the efforts of the authors of this book with a few co-workers, because their empirical knowledge is too limited for the application of this approach to various branches of scientific activities. Its success will depend on those scientists who are interested and willing to approach their subject according to the requirements presented in this book and to look at their expert knowledge in this new light.

SUBJECT INDEX

Hydrophilic Solvents, 117 ff, 215, 218
Hydrophobic Soltues 69 ff, 99 ff, 104, 189 f, 210
Hydrophobic Solvents 97 ff
Hypothesis 3

I

Idea 49
Ideal Crystal 41, 48 f, 175, 209
Ideal Gas 41
Identitas 169 f, 181
Imagination 49, 171, 176
Individuality 40, 206
Induction 181
Information 3, 189, 207 ff, 229
    Theory 208
Intellect 48
Intelligence 181, 203, 231
"Intelligent Behaviour" 36 f, 174, 231
Interacting Forces 23
Interactions
    Terminal Effects 34 f
Interface 80 ff, 89, 187 ff, 201
Intermolecular
    Distance 28 ff, 127
    Interactions 27 ff
Intimate Ion Pairs 161
Intracellular
    Structures 92
    Water 77, 209 ff
Intramolecular
    Distances 28 ff, 127
    Forces 23
    Organization 223 ff
      Haemoglobin 229
      Phenanthroline-iron Complexes 225 ff
      Solvatochromic Complexes 227 ff
      Water 230 f
Investigator 178
Ionic Theory 5

Ionization 26, 136, 157, 161
Ionotropism 5
Ion Pairs 160 ff
Iodine Monochloride 217
Irreversibility 4, 180
Isoequilibrium Temperature 229
Isokinetic Relationship 149, 229
Isosolvation Point 153

K

Kinetcs 6, 190

L

Lewis Acid 130, 160, 172
Lewis Base 130
Lewis Concept 21
LFER 132
Lipids 90, 93, 210
Liquid 39 ff
    Actions of Gases 189
    Classification 42 f
    Existential Requirements 71, 109
    Illustration of System Organization 215
    Inner Surfaces 188
    Junction Potential 81, 145
    Oscillating Pattern 189
    Quality 189
    Stability Range 59 ff, 85, 187, 193
    State 39 ff, 186 ff
    Statistical Description 41
    Theoretical Approaches 41 f

M

"Magic Acids" 160
Mathematical Description 3 f, 5, 8, 25, 170
Measurability 176
Membranes 86, 92 ff, 205, 209
Metastable State 94

of Water 57 f
of Water-Alcohol Mixtures 124
Vitrified Water 192, 196
Vocabulary in Science 178
Volume of Mixing 121

W

Water 1, 43 ff, 53 ff, 69 ff, 79 ff,
    89 ff, 97 ff, 109 ff, 117 ff,
    185 ff, 195 ff, 205 ff, 230 f,
    Anomalies 53 ff, 58 ff
    Association 44 ff,
    Compressibility 57, 197, 222
    Critical Point 65, 79
    Density 55, 74, 82, 195, 221
    Diffusionally Averaged Structure
    47, 209
    Dymeric Molecules 44
    Dielectric Constant 58
    - and DNA 85, 212 f
    Extracellular - 77, 86, 203 ff
    Existential Requirements 79,
    109, 185 ff, 188 f, 196
    Gases in Solution 98, 122, 126,
    186, 195, 221
    Heat Capacity 55, 83, 185, 189,
    195, 197, 221
    Heat of Vaporization 54
    Hierarchic Levels
        in the Liquid 180, 185, 193,
        221
        in the Molecules 203 f
    - in the Human Body 77, 205 ff
    Information Content 207 ff
    Intracellular - 77, 209 ff
    Intramolecular Organization
    203 f
    Ionization 27
    - "Language" 207, 209
    Liquid Range 53, 213
    Molecular Approach 43, 47
    Molecular Structure 43
    Mixtures with Alcohols 119 ff

- in Non-Aqueous Solvents 131,
215 ff
Organization 173, 185 ff, 230 f
Phase Boundaries of - 79 ff,
205, 209
Phase Transitions 55 f
Self-Ionization 5, 44, 72, 97,
109
Small Drops 87, 100, 187, 197
Stability Range 59 ff, 85, 187,
193
Structure 6, 43, 53 f, 197, 209
    Models 45
    Loosening 62, 102, 109, 190,
    197, 220
    Tightening 111, 113, 190, 200
Supercooled State 29, 58 ff,
126, 196 ff
Superheated State 29, 65
Surface Tension 54, 79, 83, 121
Thermal Conductivity 57, 195,
221
Thin Layers of - 60, 82 f, 187,
199
Trimeric Units 45
and the Unity of the Body 179,
207
Vapour Pressure 104, 221
Viscosity 57 f, 221
Vitrified - 192
Wave Function 14

Y

Y-values 131

Z

Zero-point Energy 15 f,
Zero-point Fluctuations 15 f
Z-values 132, 135

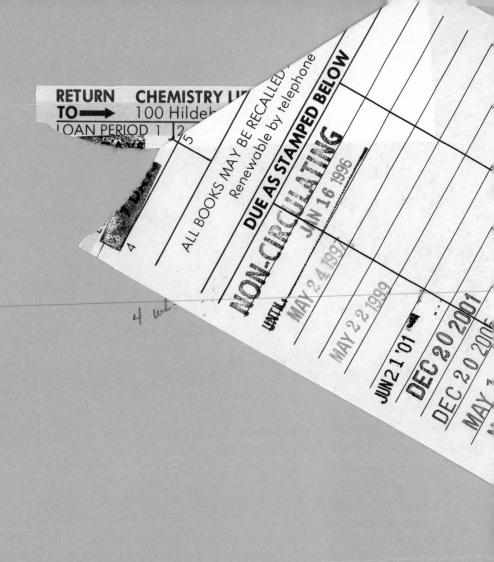